The Limits

KIT FINE

CLARENDON PRESS · OXFORD

OXFORD
UNIVERSITY PRESS

Great Clarendon Street, Oxford ox2 6DP

Oxford University Press is a department of the University of Oxford.
It furthers the University's objective of excellence in research, scholarship,
and education by publishing worldwide in

Oxford New York

Auckland Bangkok Buenos Aires Cape Town Chennai
Dar es Salaam Delhi Hong Kong Istanbul Karachi Kolkata
Kuala Lumpur Madrid Melbourne Mexico City Mumbai Nairobi
São Paulo Shanghai Singapore Taipei Tokyo Toronto

with an associated company in Berlin

Oxford is a registered trade mark of Oxford University Press
in the UK and in certain other countries

Published in the United States
By Oxford University Press Inc., New York

© Kit Fine 1998, 2002

The moral rights of the author have been asserted

Database right Oxford University Press (maker)

First published 2002
First published in paperback 2008

All rights reserved. No part of this publication may be reproduced,
stored in a retrieval system, or transmitted, in any form or by any means,
without the prior permission in writing of Oxford University Press,
or as expressly permitted by law, or under terms agreed with the appropriate
reprographics rights organization. Enquiries concerning reproduction
outside the scope of the above should be sent to the Rights Department,
Oxford University Press, at the address above

You must not circulate this book in any other binding or cover
and you must impose this same condition on any acquirer

British Library Cataloguing in Publication Data

Data available

Library of Congress Cataloging in Publication Data

Fine, Kit.
The limits of abstraction / Kit Fine.
p. cm.
Includes bibliographical references and index.
1. Mathematics–Philosophy. 2. Abstraction. I. Title
QA8.4 .F56 2002   510′.1–dc21   2002070134

ISBN 978-0-19-924618-2 (Hbk.)
978-0-19-953363-3 (Pbk.)

1 3 5 7 9 10 8 6 4 2

Typeset in 10.5 on 12 pt Minion
by SPI Publisher Services Ltd, Pondicherry, India
Printed in Great Britain by
Biddles Ltd., King's Lynn, Norfolk

THE LIMITS OF ABSTRACTION

What is abstraction? To what extent can it account for the existence and identity of abstract objects? And to what extent can it be used as a foundation for mathematics? Kit Fine provides rigorous and systematic answers to these questions along the lines proposed by Frege, in a book concerned both with the technical development of the subject and with its philosophical underpinnings.

Fine proposes an account of what it is for a principle of abstraction to be acceptable, and these acceptable principles are exactly characterized. A formal theory of abstraction is developed and shown to be capable of providing a foundation for both arithmetic and analysis. Fine argues that the usual attempts to see principles of abstraction as forms of stipulative definition have been largely unsuccessful but there may be other, more promising, ways of vindicating the various forms of contextual definition.

The Limits of Abstraction breaks new ground both technically and philosophically, and is essential reading for all those working on the philosophy of mathematics.

'The text is essential reading for anyone interested not only in abstractionist philosophies of mathematics, but the philosophy of mathematics and language in general. The philosophical chapters display a consistently high level of rigour and insight ... the new philosophical problems raised are valuable and thought provoking, and promise to be the basis for much philosophical discussion to come.'

Roy Cook and Philip Ebert, *British Journal of Philosophy of Science*

Kit Fine is Professor of Philosophy at New York University.

Preface

THIS monograph has its genesis in a paper of the same name, written in 1994 as a contribution to the proceedings of a conference on the philosophy of mathematics that was held in Munich during the preceding year. Because of various delays, these proceedings (*The Philosophy of Mathematics Today*, edited by M. Schirn) were not themselves published until 1998. Peter Momtchiloff from Clarendon Press offered to publish an expanded version of the paper as a separate monograph; and I was happy to agree. I have corrected various errors in the original paper, improved the exposition here and there, and incorporated some brief comments on the more recent literature. The major change is the addition of a new part on the context principle (which was omitted from the original paper for lack of space).

The earlier discussion of the context principle contained both a negative part, dealing with the difficulties in providing a proper formulation of the principle, and a positive part, which attempted to show how these difficulties might be met. I now appreciate that the positive part calls for a new approach to the philosophy of mathematics—what I call 'procedural postulationalism'—and that discussion of it is best postponed to another occasion. I have therefore presented only the criticisms from the negative part. But it is important to bear in mind that these criticisms are intended as the prolegomena to a more constructive account.

I am much indebted to the participants of the Munich conference—and especially to Boolos, Clark, Hale, Heck, and Wright—for reawakening my interest in the topic of logicism. Preliminary versions of the paper were given at the third Austrian philosophy conference in Salzburg, at a talk at the City University of New York, at a philosophy of mathematics workshop at the University of California at Los Angeles, and at a workshop on abstraction in St Andrews; and I am grateful for the comments that I received at those meetings. I have been greatly influenced by the writings of Michael Dummett and Crispin Wright and have greatly benefited from the comments of

Tony Martin. Joshua Schechter read through the original published paper and suggested many helpful improvements, both typographic and substantive; and Sylvia Jaffrey, for OUP, provided careful copy-editing of a disorderly text.

I am very grateful to John Burgess, Roy Cook, Philip Ebert, Stewart Shapiro and Alan Weir for pointing out some infelicities and errors in the original hardback edition of the book and I have attempted to correct these (along with some other minor infelicites) in the present paperback edition.

Contents

Introduction   ix

I. Philosophical Introduction   1

1. Truth   3
2. Definition   15
3. Reconceptualization   35
4. Foundations   41
5. The Identity of Abstracts   46

II. The Context Principle   55

1. What is the Context Principle?   56
2. Completeness   60
3. The Caesar Problem   68
4. Referential Determinacy   77
5. Predicativity   81
6. The Possible Predicative Content of Hume's Law   90

III. The Analysis of Acceptability   101

1. Language and Logic   101
2. Models   105
3. Preliminary Results   107
4. Tenability   114
5. Generation   118
6. Categoricity   122
7. Invariance   138
8. Hyperinflation   156
9. Internalized Proofs   161

IV. The General Theory of Abstraction 165

1. The Systems 165
2. Semantics 175
3. Derivations 189
4. Further Work 191

References 193

Main Index 197

Index of First Occurrence of Formal Symbols and Definitions 200

Introduction

THE present monograph has been written more from a sense of curiosity than commitment. I was fortunate enough to attend the Munich Conference on the Philosophy of Mathematics in the Summer of 94 and to overhear a discussion of recent work on Frege's approach to the foundations of mathematics. This led me to investigate certain technical problems connected with the approach; and these led me, in their turn, to reflect on certain philosophical aspects of the subject. I was concerned to see to what extent a Fregean theory of abstraction could be developed and used as a foundation for mathematics and to place the development of such a theory within a general framework for dealing with questions of abstraction. To my surprise, I discovered that there was a very natural way to develop a Fregean theory of abstraction and that such a theory could be used to provide a basis for both arithmetic and analysis. Given the context principle, the logicist might then argue that the theory was capable of yielding a philosophical foundation for mathematics, one that could account both for our reference to various mathematical objects and for our knowledge of various mathematical truths. I myself am doubtful whether the theory can legitimately be put to this use. But, all the same, there is surely considerable intrinsic interest in seeing how the theory of abstraction might be developed and whether it might be capable of embedding a significant portion of mathematics, even if the theory itself is in need of further foundation.

The monograph is in four parts. The first is devoted to philosophical matters and serves to explain the motivation for the technical work and its significance. It is centred on three main questions: What are the correct principles of abstraction? In what sense do they serve to define the abstracts with which they deal? To what extent can they provide a foundation for mathematics? The second part (omitted from the original paper) discusses the context principle, both as a general basis for setting up contextual definitions and in its particular application to numbers. The third part proposes and investigates a set of necessary and sufficient conditions for an abstraction principle to be acceptable. The acceptable principles, according to this criterion, are precisely determined and it is shown, in particular, that there is a strongest such principle. The fourth and final part attempts to develop a general theory of abstraction within the technical limitations set out by the third part; the theory is equipped with a natural class of models; and it is shown to provide a foundation for both arithmetic and analysis.

I

Philosophical Introduction

AN abstraction principle associates objects with the items from a given domain, the objects associated with two items being the same when the items are suitably related and the objects being distinct when the items are not so related. For example: we may abstract on concepts in accordance with the principle that concepts between which there exists a one–one correspondence are to be associated with the same number; or we may abstract on lines in accordance with the principle that parallel lines are to be associated with the same direction.

We shall follow Frege (1892) in taking there to be a basic distinction between objects and concepts. Objects are referred to by means of singular terms and concepts by means of predicates; and variables for objects and concepts are respectively taken to occupy either a nominal or a predicative position. Although concepts may *correspond* to objects, no concept can sensibly be said to be an object, since this would involve a grammatical confusion between a singular term and a predicate.

A principle of abstraction is said to be *conceptual* when the items upon which it abstracts are concepts and it is said to be *objectual* when the items upon which it abstracts are objects. Thus of the two examples above, the first is a principle of conceptual abstraction and the second a principle of objectual abstraction. The two kinds of principle are fundamentally different; for the conceptual principles involve a 'projection' of the larger domain of concepts into the smaller domain of objects, where these objects may themselves fall under the concepts upon which an abstraction is made. It is this special reflexive feature of the conceptual principles that makes them so powerful—and also so dangerous.

The serious study of conceptual abstraction, as it is understood here, began with Frege. He was the first to provide a clear statement of its principles and the first to provide a cogent account of how such principles might be of use in determining an ontology of abstract objects and a foundation for mathematics. Two principles of conceptual abstraction are prominent in his work. The first is the principle for abstracting to numbers, which was mentioned above and will be called 'Hume's Law' or 'Principle'. The second is the principle of extensional abstraction, Law V of the *Grundgesetze*; it associates extensions with concepts, the extensions being the same when the objects falling under the concepts are the same.

Frege attempted to derive the whole of arithmetic and analysis from Law V with the help of other, less problematic, logical principles. But the attempt failed if for no other reason than that Law V leads, by the argument of Russell's paradox, to a contradiction. One might have thought, prior to the discovery of the paradoxes, that extensional abstraction was innocuous. For what is to prevent one from picking out the extension of each concept? Indeed, the appeal of extensional abstraction in this regard would seem to extend to all forms of conceptual abstraction. For just as there would appear to be nothing to prevent one from picking out the extension of each concept, so there would appear to be nothing to prevent one from picking out its counter-extension or its number or any other aspect of it.

But the appeal of conceptual abstraction in this regard depends upon overlooking the critical respect in which it differs from objectual abstraction. For what makes it appear so innocuous is the view of concepts as being given independently of the objects that are abstracted from them. Once it is recognized that the abstracts themselves may play a role in the determination of concepts, it is no longer so clear that one can simultaneously abstract on the concepts and 'conceptualize' over the resulting objects.

Russell's paradox has, until recently, tended to deflect attention away from the topic of conceptual abstraction. Philosophers and logicians alike have followed Frege's own lead in considering other approaches to the foundation of mathematics and have not been so interested in the topic of conceptual abstraction as such. The prevalent view is that abstracts should just be treated as equivalence classes (or perhaps we should say equivalence *sets*). The theory of abstraction thereby becomes a part of the much more comprehensive theory of sets or classes.

But conceptual abstraction, as so conceived, represents a double departure from the Fregean conception. In the first place, it is not conceptual, i.e. it is not abstraction on concepts, since the members of the classes are objects, not concepts; in the second place, it is not even Fregean abstraction, since the classes in question will not arise from a principle of abstraction of the sort envisaged by Frege.

Maybe the Fregean idea of conceptual abstraction should be given up; and maybe the current conception of abstraction should take its place. But it is surely premature to dismiss the Fregean approach on the grounds of a single failure.

In developing the Fregean approach, there are three main questions that need to be considered. The first is which, if any, of the principles of conceptual abstraction are true (or otherwise acceptable); the second is, given that an abstraction principle is true, what kind of truth is it and how, in particular, does it relate to the objects with which it deals; and the third is whether abstraction principles can serve as a foundation for mathematics or, at least, for a significant part thereof. Frege, in his state of pre-Russellian innocence, could provide oversimplified answers to all three questions. He could embrace all abstraction principles indiscriminately by way of their reduction to extensional abstraction; he could treat Law V as definitive of extension; and, with the help of the Law, he could provide a foundation for arithmetic and analysis. On all these matters, he was mistaken. But we, in our state of post-Russellian sophistication, can still profitably consider the questions that Frege's work raises even if we cannot accept his answers.

1. Truth

Of our three questions the most important is certainly the first; for it is only when one has settled on the truth of a principle of abstraction that one can sensibly raise the question of its status as a truth or its role in providing a foundation for mathematics. Any abstraction principle can be stated in the form:

the abstract of α = the abstract of β iff $\ldots \alpha \ldots \beta \ldots$,

where the variables 'α' and 'β' range over the items (be they objects or concepts) on which the abstraction is to be performed, the phrase 'the abstract of' stands in for an abstraction operator, such as 'the number of' or 'the direction of', and the clause '$\ldots \alpha \ldots \beta \ldots$' represents a criterion of identity, such as equinumerosity or coextensiveness. Two necessary conditions for the truth of an abstraction principle hold as a matter of logic (see sect. 3.4). In the first place, it follows from the truth of an abstraction principle that its underlying criterion of identity on concepts should be an equivalence relation (reflexive, symmetric, and transitive). Thus each abstraction principle will induce a partition of the domain of items into mutually exclusive equivalence classes.

Secondly, it follows from the truth of an abstraction principle that the identity criterion should not be inflationary, the number of equivalence classes must not outstrip the number of objects. There must, that is to say, be a one–one correspondence between all the equivalence classes, or their representatives, on the one hand, and some or all of the objects, on the other hand. It is, of course, exactly on this score, that Law V proves unacceptable; for where there are n objects, it demands that there be 2^n abstracts.

The question naturally arises as to whether these two necessary conditions are jointly sufficient for the truth of a principle of conceptual abstraction. In considering this question, it is important to distinguish between circular and non-circular criteria of identity. A criterion is circular or not according as to whether it involves the very notion of abstraction that is in question.[1] The standard criteria of identity, such as equinumerosity or coextensionality, are non-circular; while the identity criterion behind Frege's proposed amendment to Law V is not, since coextensionality is restricted to the objects that are distinct from the given abstracts.

It is clear from the consideration of examples that our proposed conditions are not sufficient for the truth of abstraction principles whose identity criterion is circular. The general difficulty is that how one associates objects with equivalence classes of concepts will help determine what those equivalence classes are. Thus there may be no consistent way of simultaneously fixing the equivalence classes and associating them with objects (see the comments to lemma III. 4.1).

In the non-circular case, however, there is no such impediment to the truth of an abstraction principle; the concepts can be associated with objects in such a way as to render the principle true (lemma III. 4.1). It is therefore tempting to suppose, on grounds of a general

---

[1] A non-circular criterion can involve the notion of abstraction associated with some other criterion of identity as long as there is no circularity in the system of principles as a whole. The possibility of such non-circular systems of principles is countenanced in the technical discussion of Parts III and IV but, for purposes of simplicity, I have here assumed that no non-circular criterion will invoke any other criteria.

principle of plenitude for abstract objects, that an abstraction principle, in such a case, is true. Since there *can* be abstract objects of the sort in question, it is supposed that there *are* such objects.

It is important in considering this view to distinguish the question of whether a given principle is correct from the question of whether it is genuinely a principle of abstraction. Whether something is a principle of abstraction has been for us a matter of form; the principle must state an equivalence between an appropropriate form of identity statement, on the one side, and an appropriate form of identity criterion, on the other. But from among the abstraction principles in this formal sense, only some are abstraction principles in a genuine or 'material' sense. This point can be made especially vivid if we bear in mind that to any statement whatever there will correspond an abstraction principle; for we may take its criterion of identity to be one which identifies two concepts just in case the statement holds (cf. Heck (1992)). The statement and the principle will then be equivalent (at least granted the existence of a universal method of abstraction, one which identifies all concepts). But a principle obtained in this way will not in general be a genuine principle of abstraction; for its truth will not rest entirely, or even principally, on the existence of abstracts.

I assume it would be very difficult to say what makes a principle a genuine principle of abstraction. However, our present concern is not with picking out the genuine principles from among those that are true but in picking out the true principles from among those that are of the required form; and, regarded in this way, the proposed condition of sufficiency is far more plausible.

Unfortunately, it is still subject to difficulties. One, which we dub 'additivity', is familiar from the literature (see e.g. Boolos (1990: 273–4)). For two principles may make incompatible demands on the size of the universe, even when each alone makes a consistent demand on its size. We may imagine a principle, for example, that inflates on finite universes but not on infinite ones and another principle that inflates on infinite universes but not on finite ones. The two principles, when taken together, will inflate on any universe; and so our proposed conditions cannot be taken to guarantee the truth of the principles.

Such a difficulty could be overcome if the domain of objects were taken to be given in advance of the adoption of any principle of abstraction. The non-inflationary principles could then be taken to be those that were non-inflationary on this particular domain rather than on some domain or other; and the clash between the different non-inflationary principles would then be avoided. The natural way to fix the domain in advance is to let it be the universal domain, i.e. to be inclusive of all objects whatever. The idea is that somehow, prior to the adoption of any principle of abstraction, we can determine the size of the universe and that, once this is done, we can then determine which of the principles are true.

But even if we allow that a universe, of fixed size, can be given prior to the adoption of any principles of abstraction, there is another, less familiar, difficulty that can arise. Let us suppose that there are more principles than objects or, to be more exact, that they give rise to more equivalence classes of concepts than there are objects. Then, on the assumption that the abstractions associated with distinct equivalence classes of concepts are distinct, there will again be more abstracts than objects. Thus although no principle alone inflates on the given universe, taken together they do inflate. We have what might be called the problem of *hyper*inflation.

This problem can arise even with abstraction principles that merely serve to divide the universe of concepts into two. For given any concept, we may divide the universe of concepts into those that are coextensive with the given concept and those that are not. Each such division then yields 2 abstracts; but, taken together, they yield 2^n abstracts, where n is the size of the universe.[2]

---

[2] There is a corresponding problem of hyperinflation for objectual abstraction even though there is no corresponding problem of inflation. For given any concept F, we may divide the universe of objects into those that F and those that do not. Thus once we allow there to be multiple principles of abstraction, the difference in safety between the objectual and conceptual cases is not so great.

Although the problem has been stated in the metalanguage using standard set-theoretic machinery, it is readily reproduced in the object-language. Thus, in the case of objectual abstraction, let [F] be the divisor method of abstraction associated with the concept F. Subject [F] to the principle:

$\forall F \forall x, y \, ([F]x = [F]y \leftrightarrow (Fx \leftrightarrow Fy));$

lay down the identity postulate:

$\forall F, G \forall x, y \, ([F]x = [G]y \rightarrow \forall z \, ((Fz \leftrightarrow Fx) \leftrightarrow (Gz \leftrightarrow Gy)));$

and define ϵ by:

$x \, \epsilon \, y =_{df} \exists F \, (y = [F]x \,\&\, Fx).$

Then we may show:

$\forall F \, (\exists x Fx \rightarrow \exists y \forall x \, (x \, \epsilon \, y \leftrightarrow Fx));$

and from this, by the reasoning of Russell's paradox, follows $\forall x, y(x = y).$

I propose to solve this problem by imposing a further condition on the non-circular criteria of identity. For some are purely logical in the sense that they can be formulated without the aid of non-logical concepts. Thus the familiar criteria of coextensiveness and of equinumerosity are logical; but the divisor criteria above (with the sole exceptions of division by the universal and empty concepts) are not. Each logical criterion corresponds to an invariant equivalence relation, one that can take no account of the specific identity of the different objects (or, more technically, to one that is invariant under any permutation).

Under a certain assumption, it can be shown that the principles, as so restricted, will not hyperinflate. Let us say that a transfinite cardinal c is *unsurpassable* if $2^d \leq c$, where d is the number of cardinals less than c. Then, by means of the analysis of invariance in sect. III.7, it may be demonstrated that the non-inflationary and logical abstraction principles will fail to hyperinflate just in case the cardinality of the domain is unsurpassable. Thus the unsurpassable cardinals are a kind of analogue, in our theory, to the inaccessible cardinals of ZF; and, just as with the inaccessibles, the question of their existence cannot be settled within ZF (sect. III.8).

The proposal can be taken further. For let us say that an abstraction principle is *predominantly logical* (or *invariant*) if its identity criterion involves only a small number of objects in relation to the number of objects in the universe as a whole.[3] For example: given a small class of objects, the agreement of two concepts with respect to the objects in the class is predominantly logical; and any criterion which uses the concept of being an individual will be predominantly logical as long as there is a small number of individuals in relation to the number of objects as a whole. The result can then be extended to the principles that are predominantly logical; they too will not hyperinflate as long as the domain is of unsurpassable cardinality.

A kind of converse result also holds. For if we allow abstractions to be defined on a large number n of objects, then hyperinflation will

---

[3] This notion of being small is not the usual one. A subset C of cardinality c said to be exponentially small relative to a domain D of cardinality d if $d^c \leq d$, i.e. if the number of subsets of the same cardinality as the given subset does not exceed the cardinality of the domain itself. A predominantly logical criterion is one whose extension is invariant under any permutation that is an identity on some exponentially small set of cardinality c for which it is also true that $2^{2^c} \leq d$.

result. For corresponding to any set of n objects, there will then be a unique abstract; and there are more such sets, and hence more such abstracts, than there are objects in the universe. Thus the restriction to those identity criteria that are predominantly invariant is, in a sense, the best possible.

It follows from the preceding considerations that the two necessary conditions, being an equivalence on concepts and being non-inflationary, can be taken to be jointly sufficient for the truth of any principle of abstraction that is non-circular and predominantly logical. Furthermore, it can with some plausibility be supposed that any principle of abstraction should be predominantly logical; for if one non-inflationary principle involving a large number of objects is permitted it is hard to see why all such principles should not also be permitted. It is also plausible to suppose that abstraction principles should be like definitions in general in being non-circular. Under these two suppositions, we can then obtain a necessary and sufficient condition for the truth of any abstraction principle whatever; for an abstraction principle will be true (within the whole universe) just in case its identity criterion is non-circular and yields a non-inflationary and predominantly logical equivalence on concepts.

The two requirements provide a *condition* of truth. But how useful are they in providing a *test* for truth? To what extent can they be employed in establishing that a given abstraction principle is true? In the special case in which an abstraction principle is non-circular, i.e. innocent of reference to the notion of abstraction, the requirements for its truth will likewise be innocent of such reference. Indeed, it will be possible to formulate the requirements within the language of higher-order logic—and, under certain simplifying assumptions, within second-order logic itself (sect. III.4). It might therefore appear that we could settle which abstraction principles are to hold on the basis of the corresponding higher-order or second-order conditions and thereby avoid any appeal to intuitions regarding the abstracts themselves.

However, the epistemological advantages that such a reduction appear to provide are largely illusory. For the satisfaction of the corresponding higher-order conditions is itself to be settled on the basis of substantive non-logical considerations. Consider Hume's Law, for example. The satisfaction of the corresponding non-inflation condition will require the existence of infinitely many objects. And how is this to be ascertained without presupposing that there are infinitely many numbers, or sets, or abstract objects of some other sort?

Still, the proposed conditions are not altogether useless as a test. Let us distinguish between objects of abstraction, which are given by a principle of abstraction, and abstract objects in a broader sense of the term. Thus points in abstract Euclidean space are abstract objects and yet are not objects of abstraction, since they are not introduced, and cannot sensibly be taken to be introduced, through a principle of abstraction of the sort envisaged by Frege. The conditions then enable us to settle the question of the existence of objects of abstraction of a given kind either on the basis of the existence of objects of abstraction of some other kind or kinds or on the basis of the existence of abstract objects that are not given through abstraction. An example of the first kind of justification is illustrated by our own axiomatic theory of abstraction; for within that theory, an infinity of divisor abstractions (ones that divide the universe of concepts into two) will enable us to justify number abstraction. A powerful example of the second kind is provided by standard set theory; for its huge ontology will bring a host of non-inflationary conditions in its wake.

Our conditions have so far been stated by appeal to an informal concept of truth. I want now to consider two model-theoretic criteria of acceptability and see how how they compare with the informal criterion. (The discussion to the end of this section is somewhat technical and may be omitted by the less technically minded reader.)

We can think of each model-theoretic criterion as being obtained from the informal criterion by adopting an appropriate model-theoretic criterion of truth. The first, which we call 'tenability', results from taking the universe of objects to be given as a set. This set will determine a standard model, one containing a concept for each subset of objects; and truth can then be taken to be truth relative to the model.

Under the simplifying assumption that the abstraction principle is logical or 'invariant', the truth of the equivalence and non-inflation conditions for that principle in a given model will depend only upon its cardinality. We may therefore say that an abstraction principle is *tenable on* a cardinal if its identity criterion determines a non-inflationary equivalence on the concepts in a standard model of that cardinality (sect. III.4).

The second criterion, which we call 'stability', is obtained by treating truth as a limit concept; a statement is taken to be true when it is true in all models of sufficiently large cardinality. Adopting this criterion, we may then say that an abstraction principle is *stable* if, for some cardinal, the principle is tenable on all greater cardinals (cf. Heck 1992: n. 4).

How faithful are the model-theoretic criteria to our informal conception of truth? In considering this question we must take account of the attitude of the abstractionist to standard set theory (as embodied in ZF or ZFC). He can either be compromising or uncompromising. The uncompromising abstractionist rejects set theory. He therefore sees the theory of abstractions as an alternative, rather than as a supplement, to the standard theory of sets.

There are, of course, various grounds upon which ZF might be rejected. But there is one directly related to the abstractionist's position. For he may adopt an imperialistic stand and see all abstract objects as arising from abstraction. For him, to be an abstract object is simply to be an object of abstraction, one that is introduced by means of an appropriate principle.

The attractions of the thoroughgoing position are manifest; for it provides us with a single and simple unified point of view from which the various philosophical problems concerning abstract objects may be considered. But the difficulties in the position are enormous. For many kinds of abstract object, from both within and without mathematics, do not appear to fit within the abstractionist mould. They must therefore be rejected or, at best, accepted in a distorted form.

What is the uncompromising abstractionist to make of the tenability or stability criteria? It appears as if they must be rejected— presumably as unintelligible, though possibly as false. For the criteria—with their reference to models, cardinals, satisfaction, and the like—are stated within the language of ZF, and so must be rejected along with ZF. Perhaps there is some way of stating the criteria, or something like them, in terms acceptable to the abstractionist. But if there is a such way, it is not at all clear what it is.

The advocate of ZF, on the other hand, is in a position to recognize the criterion as a correct account of truth for the uncompromising abstractionist. For he can acknowledge that, for certain cardinals, the principles tenable on those cardinals might be exactly those that would be true were his opponent correct. We have therefore what might be called an *externalist* characterization of a given position, one which can be regarded as correct, or even as intelligible, only by someone who does not hold the position. (A similar and more familiar case is provided by the classical characterizations of the constructivist conception of logical validity.)

Given that we adopt the externalist perspective, it is natural to ask: What is the size of our opponent's universe? In response to this question, it seems that the best one can do is to see which cardinals will render true what our opponent takes to be true. Thus if he adopts the principle of plenitude, as stated above, then we will know that the cardinal will be unsurpassable and hence uncountable. But this will not enable us to distinguish between his acceptance of one unsurpassable as oppose to another.

In the face of this difficulty, one might be tempted to suppose that our abstractionist should endorse any abstraction principle that stabilizes on some cardinal. For if his thinking is governed by a general principle of plenitude, then should he not recognize as true anything that would be true in a sufficiently large universe? But this is to import too much of the externalist standpoint into his own way of thinking. For how large can a universe be? For us, the possible size of the universe can be determined on the basis of set theory. But, for him, it must be determined in some other way; and if he is thoroughgoing, it must be determined on the basis of the methods of abstraction themselves. There is no way of seeing 'from above' which methods of abstraction can be accepted and which not.

Consider, for example, the restricted forms of the principle of extensional abstraction. They allow one to form the extension of a concept that holds of a sufficiently small number of objects—finitely many, or countably many, or what have you. We would know that such a principle was non-inflationary if the universe were sufficiently large. But from the standpoint of a thoroughgoing abstractionist, this is not something that can simply be assumed. It must itself be established on the basis of principles that are independently plausible. In fact, I shall later suggest that there are such principles and that they enable us to determine the cardinality of the universe of the uncompromising abstractionist as being that of the first unsurpassable.

We now turn to the case of the abstractionist who is prepared to endorse set theory. In considering this case, it is important to bear in mind that the proper setting for the theory of conceptual abstraction is second-order logic; for it is this that provides us with the logic of the concepts upon which the abstraction is made. We shall suppose that the second-order logic employed contains a general comprehension axiom, one to the effect that each condition determines a concept. It then follows, in particular, that there is a universal concept, one holding of every object in the domain. We must now ask: What is the attitude of our advocate of set theory to second-order logic? How does he conceive of the domain of objects from which the concepts are to be drawn? Two basically different attitudes may be distinguished. He might, on the one hand, be prepared to accept second-order logic as applying to the whole universe of sets and urelements. In this case, he will be willing to endorse the second-order version of ZF, with concepts having a far wider range of application than sets.

On the other hand, he might think of the domain of objects from which the concepts are drawn as itself a set. Indeed, this attitude is forced upon him if he thinks of sets as the most general method of collection; for given that there were a universal concept, one holding of every object in the underlying domain, there would have to be a universal set, one to which every object in that domain belongs. In this case, our theorist will merely endorse first-order ZF and will most naturally think of second-order logic as being interpreted within it. Thus, in the one case, our theorist will think of second-order logic as lying *alongside* ZF; while, in the other, he will think of it as lying *within* ZF.

Consider the second case first. Since there is no fixed domain for second-order logic, there is no fixed domain for the theory of abstraction and hence no fixed conception of truth. The proper setting for a theory of abstraction on such a view is not really second-order logic but the first-order theory of sets. Thus the resulting theory will be of a quite different sort from the one envisaged by Frege.

All the same, we may ask: Which second-order principles of abstraction can be made true on a domain of given cardinality? And presumably, the answer is: those principles that are tenable on the given cardinal. For it seems reasonable to suppose that the resulting abstracts exist; and given that they exist, they can be included within the domain without danger of inflation.

But it would also be desirable to have an absolute criterion of acceptability, one that is not relative to a given domain or cardinal. We would like the acceptable principles to be true in domains of arbitrarily large size, so that they can be applied regardless of the size of the underlying domain; and we would like singly acceptable principles to be jointly acceptable, so that the application of one such principle does not exclude the application of another. We are therefore led to propose the following requirement: all acceptable principles should be jointly tenable on arbitrarily large domains.

In attempting to meet this requirement, it will not do to *define* acceptability as tenability on arbitrarily large cardinals. For one principle may be tenable on exactly the successor cardinals while the other may be tenable on exactly the limit cardinals. Both will therefore be acceptable according to the criterion; and yet they will not be jointly tenable.

A natural solution to this problem is to use stability as the criterion of acceptability. For any set of stable principles can be made jointly true in a domain of sufficiently large cardinality (even allowing that different principles invariably yield different abstracts). But the solution is not altogether satisfactory. For a proper class of stable principles cannot, in general, be rendered jointly true. Given any cardinal, for example, we can 'rig' an abstraction principle that is untenable up to that cardinal and tenable thereafter. But then no set-domain can render all such principles true.

Nor is the proposal optimal (under the more limited requirement that any *set* of acceptable principles can be made jointly true). For the requirement may be satisfied by taking a principle to be acceptable when the class of cardinals upon which the principle is tenable contains an unbounded subclass of cardinals that is closed under limits.[4] Indeed, it is not clear that there is an optimal solution; and even if there is, it is far from being unique.[5]

We turn, finally, to the case of a compromising abstractionist, one who is prepared to endorse second-order set theory. It might appear from our previous discussion that such a position is impossible. For

---

[4] I owe this suggestion to Tony Martin. But note that even this proposal may not be optimal; for in a model of set theory whose cardinality is weakly Mahlo, the class of regular cardinals will be stationary and hence can be added to the given acceptability—inducing classes without violation of the requirement.

[5] Assume that the set-theoretic universe is of cardinality c. A 'solution' will then correspond to a family of sets of cardinals $<$ c where, intuitively, these are the sets of cardinals upon which an acceptable principle may be tenable; and an optimal solution will be one in which the family is a c-additive non-principal ultrafilter. Thus an optimal solution will only exist if the cardinal c is measurable; and given that an optimal solution exists, we may use arbitrary permutations on the domain of cardinals to obtain at least c other optimal solutions.

set theory requires a domain (if we were to think of it as a set) whose cardinality is inaccessible, whereas abstraction theory, under a principle of plenitude, requires a domain whose cardinality is unsurpassable. But no inaccessible cardinal can be unsurpassable; and it therefore appears as if no set-domain can be taken to provide a model both for set theory (i.e. ZFC) and for our theory of abstraction.

But this line of thought is in error, even granted that it is appropriate to conceive of the domains as sets. For the proper framework for a joint theory is ZFI, i.e. ZF (or ZFC) with urelements. The abstracts can then figure among the urelements; and there is nothing to prevent the domain of urelements being of greater cardinality than the domain of pure sets.[6]

What can be shown to be impossible, though, is the non-circular representation of abstracts by means of sets, one in which the abstracts are not themselves used in the sets by which they are represented. For within second-order ZFI with abstracts, there will be a principle of abstraction corresponding to any condition on cardinals. Each such condition will yield a 'generalized cardinal' that is the abstract of the concepts whose extension is of a cardinality that conforms to the condition. It may then be proved, within the theory itself, that there is no one–one correspondence between these abstracts and the pure sets (Theorem IV. 1.2). But any non-circular method of representing abstracts by means of sets should yield such a one–one correspondence; and so no such method is possible. The means of forming abstracts are therefore seen to extend beyond the reach of set theory.

We have here the ultimate *coup de grâce* for the identificatory standpoint. It has been customary to identify abstracts with equivalence classes. But sometimes, as with the cardinal numbers, this has not been possible; and the abstracts have then been identified with representatives from the equivalence classes. But the present result shows that, even allowing for arbitrary methods of identification, it will not be possible to find a place for every abstract within the set-theoretic universe. Abstracts, on grounds of cardinality alone, must be treated as objects in their own right.

What attitude, then, should the compromising abstractionist adopt towards the two formal criteria? The situation here is not so very different from that which prevails in set theory. We can hope that

---

[6] Such a possibility is considered by Menzel (1986).

some cardinal will reflect the behaviour of the whole universe, with the principles tenable on that cardinal exactly coinciding with those that are true. But if there are such cardinals, we cannot, in the light of the semantic paradoxes, expect to be able to identify them independently of the concept of truth (since otherwise we could provide a definition of truth for set theory within set theory itself).

We might therefore attempt to *approximate* to the truth. One rather crude way is to appeal to the notion of stability. Thus, granted that any true abstraction principle is true in domains of arbitrarily large size, stability will provide a sufficient condition for truth. But a more sophisticated way is to identify cardinals that reflect more and more of the properties that we perceive the universe to have. It will be natural on the generative conception of abstractions developed below to suppose that when the universe of pure sets has cardinality c the universe of abstracts will have as its cardinality the smallest unsurpassable cardinal greater than c. In this case, the present problem will reduce to the analogous problem in set theory of identifying the cardinals that reflect more and more of the properties of the universe of pure sets.

2. Definition

It is natural to suppose that we can gain an understanding of number through Hume's Law, that it can somehow serve to say what number is. In the same way, it might be thought that the abstraction principle for directions can serve in defining direction and that, in general, any acceptable abstraction principle can serve in defining the sort of abstract with which it deals.

My aim in this section is to examine some ways in which Hume's Law might help in defining number or in which an abstraction principle might help in defining some particular kind of abstract. The focus in the present section is on definitions of an orthodox kind. Later, in the next section and in Part II we shall look at definitions that are unorthodox in one or another respect.

It will be helpful, before considering Hume's Law itself, to outline what I take to be the orthodox conception of definition. A definition, we shall suppose, is of something linguistic by means of something linguistic. That which is defined is called the 'definiendum' and that by which it is defined the 'definiens'.

We may also talk of defining a non-linguistic item by means of something linguistic. We may say, for example, that 'the successor of 1' is a definition of the *number* 2 rather than of the *numeral* '2'. But in such cases we define the object by defining, or by providing the means for defining, an expression for the object. We shall later have occasion to consider cases of non-linguistic definition that are not straightforwardly reducible, in such a way, to the linguistic case.

We may distinguish between explicit and implicit definition. In an explicit definition, the definiendum will be a term (of a given grammatical type) and the definiens will be a term of the same grammatical type. The aim of such a definition is then to set up an appropriate equivalence between the terms. To take two familiar examples of this sort: 'bachelor' may be defined as 'unmarried man'; or 'God' as 'the being no greater than which can be conceived'. In an implicit definition, on the other hand, the definiens will be a statement or condition that involves the term to be defined. Thus '+' may be defined by the condition: $n + 0 = 0 \ \& \ n + m' = (n + m)'$. The purpose of such a definition is not, of course, to set up an equivalence between the term and the defining condition but somehow to constrain the interpretation of the term by means of the condition.

We shall find it convenient to treat the explicit form of definition as a special case of the implicit form. Thus the explicit definition of 'bachelor' above will be treated as equivalent to the implicit definition of 'bachelor' by means of the condition, 'All and only bachelors are unmarried men'; and, similarly, the explicit definition of 'God' will be treated as equivalent to its implicit definition by means of the condition 'God = the being no greater than which can be conceived'.

An implicit definition may be simultaneously of several terms; and the terms, whether there be one or many, may be defined by several conditions. These conditions can perhaps be regarded as one by forming their (possibly infinite) conjunction; and the terms can perhaps be regarded as one by refashioning the definition to be of a term for a sequence and then defining the original terms as the members of the sequence. Be that as it may, our focus will be on the cases in which there is but a single term and a single condition; and we shall only consider the complications of there being several terms or conditions as they arise.

The purpose of a definition is to assign an interpretation to a hitherto uninterpreted expression. We may take the interpretation to consist either in the assignment of a referent or of a sense or both.

Thus prior to the definition, the definiendum may be taken to be without a referent or sense. After the definition, if it is successful, the expression will have been assigned a referent or sense. One may, of course, talk about defining locutions that are already understood; and it is presumably in this way that Hume's Law serves to define number. But what we are saying, in such cases, is that if the expression were devoid of meaning, then the interpretation assigned by the definition would match, or correspond to, the interpretation that it already has.[7]

Let us say that a definition has *referential import* if it is meant to result in the assignment of a referent to the term t that is to be defined. The assignment of the referent should, of course, be in conformity with the defining condition D(t). It is characteristic of what I have called the orthodox conception of definition that conformity amount to no more than the requirement that D(t) should be true; the reference of t is simply to be determined by the truth of D(t). When a definition with referential import is successful, i.e. when it succeeds in appropriately assigning a referent to its term in conformity with its defining condition, we shall say that it is *referentially effective*.

A definition may result in the assignment of more than one referent to its term—not in the sense that there is the simultaneous assignment of several referents to the term, but in the sense that there are several assignments, each of a single referent, to the term. A definition with referential import is said to be *deterministic* if it is meant to result in the single assignment of a referent; and it is said to be *indeterministic* if it is meant to result in an assignment of a referent but not necessarily a single assignment. For a deterministic definition to be referentially effective it must succeeed in assigning a single referent to its term and for an indeterministic definition to be referentially effective it must succeed in assigning at least one referent to its term.[8]

---

[7] We may also take a definition to *redefine* a previously interpreted expression. In some cases (an example is the assignment of $n + 1$ to 'n' in certain programming language), a previously interpreted expression may be used in setting up a new interpretation for that very expression.

[8] The definition of a constant, such as 'Let 2 be the successor of 1', is naturally taken to be deterministic, while the definition of a variable, such as 'Let n be a number', is naturally taken to be indeterministic. In certain cases, it may be unclear whether what is being defined is a variable or a constant. I have made it a requirement on the successful definition of a variable that it should be assigned at least one value. But this requirement might be dropped.

There are semantical analogues to these referential notions of definition. We may say that a definition has *semantical import* if it is intended to result in the assignment of a sense to the term to be defined, where the sense is to be in conformity with the defining condition. Likewise, we may draw a distinction between the determinate and indeterminate forms of definition. In the semantical case, however, we will not wish to think of indeterminacy as a form of ambiguity, i.e. in the assignment of different senses to the term to be defined. The distinction must therefore be drawn along somewhat different lines.

Accordingly, we may say that a definition with semantical import is *deterministic* if it is meant to assign a sense that purports to be of a single object and that it is *indeterministic* if it is meant to assign a sense that purports to be of a range of objects. Thus the definition of 'God' by the condition 'God = the being no greater than which can be conceived' will be deterministic, while the definition of 'n' by the condition 'n is a number' will be indeterministic.

But what should the sense be taken to be in such cases? What is it for the sense of the term t, whether it be of a deterministic or indeterministic sort, to be 'in conformity' with the defining condition $D(t)$? Our answer to the corresponding question concerning reference makes natural an answer to the question concerning sense. When the definition is deterministic, the sense assigned to the term t should be the sense of the description 'the x such that $D(x)$'; and when the definition is indeterministic, the sense assigned to the term t should be the sense of a variable whose range is determined by the predicate $D(x)$.[9] (I leave open the question of how the idea of the sense of a variable should be made to fit in with the rest of Frege's views on sense.) Given that the sense is assigned in this manner, the referents assigned by the definition will be just those determined by the sense.[10]

---

[9] There are certain cases in which we might want to give a somewhat different account of the sense. Consider the definition of 'God' by the condition 'God = the being no greater than which can be conceived'. Then rather than take the sense of 'God' to be that of 'the x such that x = the being no greater than which can be conceived', we might take it to be the sense of the embedded description 'the being no greater than which can be conceived'. In this way, the sense determined by the implicit definition could be made to be the same as the sense determined by the corresponding explicit definition.

[10] Horwich (1998: 132–3) takes the sense to be determined in strict analogy to the reference. Thus the sense of t in the definition $D(t)$ will be the sense that it must have in order for $D(t)$ to be true. But then even an explicit definition—such as (x) (x is a bachelor

We see that there are four different aims a definition might have: it might attempt to assign either a referent or a sense to the term to be defined; and in each case, it might attempt to do so either deterministicly or indeterministicly. On a Fregean view, at least, it is impossible for a definition to assign a referent without also assigning a sense. Any definition with referential import should therefore be taken to have semantical import. However, not all definitions need assign a referent or even be taken to have referential import. The definition of 'God' as 'the being no greater than which can be conceived', for example, might merely be intended to fix the sense of 'God'; and it will not have a referent unless, of course, there is such a being.

On the other hand, even if a definition fails to assign a referent, it will always yield a sense; for whatever the defining condition $D(t)$, there will always be a sense corresponding to the sense of the description 'the x such that $D(x)$' or the sense of the variable whose range is determined by $D(x)$. Indeed, on a Fregean conception of language, it is hard to see how a definition could serve its purpose without assigning a sense to the term or terms that are to be defined.

In the light of these general remarks, let us now consider the possible definitional role of Hume's Law and other such principles. Given that Hume's Law is to be construed as a definition, then what should it be taken to define?

The most natural answer is that it defines the operator 'the number of'. Indeed, if one takes the syntax of the Law at its face value, then this would appear to be the only possible answer. For the other terms occurring in the Law are all of a logical character; and it would clearly be absurd to suppose that the Law somehow served to define them by means of the other terms.

There is a way, however, in which the Law (or whatever principle is in question) might be taken to do more. Any definition of a term, such as the number operator, presupposes a domain of discourse over which the term is to have application. Now under the standard forms of definition this domain is taken to be given along with the meaning or interpretation of the undefined terms. The purpose of the definition is then to assign an interpretation to the undefined term from within the given domain. Consider, for example, the recursive

---

iff x is an unmarried man), where t is the predicate 'is a bachelor'—will not yield a unique sense, since any sense that delivers the right extension will render the sentence true.

definition of addition as given by the equations: $m + 0 = 0$; and $m + n' = (m + n)'$. It is here understood that the domain of discourse is to be the set of natural numbers; and it is not required that the definition supply an interpretation of the symbol '+' over the rationals or the reals or some other kind of number or object.

Now one might adopt a similar approach in the case of abstraction. One might take the intended domain to be the universe of all objects and let the number operator have application to the concepts over that domain and the objects within it. However, even from this standpoint, it would be natural to suppose that the principle is capable of securing the interpretation of the operator for various other domains. For the success of the principle in determining an interpretation of the operator should not depend upon the individuals being what they are; and so the principle should also have application to those universes that would arise under a different disposition of what individuals there are.

But one might be more ambitious still and take the principle itself to be of help in determining the different domains of discourse. We no longer regard the domains of objects from which the values for the abstraction operation are to be drawn as simply given. Rather, we think of the principle itself as helping to determine what those domains should be. It is somehow meant to succeed both in telling us what objects there are and in assigning them to the concepts.

We may be more precise. Before we only required that the principle determine an operation F for each appropriate choice of the domain M. Now we might require the principle should determine M on the basis of the underlying subdomain I of individuals (or non-abstracts). Thus the principle must not only determine the operator F as a function of M, it must also determine M as a function of I. It must, that is to say, also determine a method of 'generation', a function g that will take us from a given domain I of individuals into an appropriate domain $M = g(I)$ of individuals and abstracts.[11]

The generation of the domain M from I may be compared to the generation of sets under the cumulative hierarchy from a given domain of urelements. Just as there are different universes of sets for different domains of urelements, so there may be different universes

---

[11] In the more general case there will be several different abstraction principles. Each principle will then be associated with a function g. The total domain (if it exists) will be obtained by successively applying the generation functions to the initial subdomain of non-abstracts until no further applications can be made.

of numbers for different domains of individuals. For different domains of individuals of finite size, the corresponding universes of numbers will be the same (in contrast to the case for sets). But it is plausible to suppose that for larger and larger domains of individuals (which may include abstracts of other kinds), new numbers of ever greater size will be added to the universe.

Now there are two rather different ways in which the Law may be taken to perform this second task. One rests upon treating the definition as having a creative capacity, as having the power to introduce new objects into the domain of discourse. This is an approach that we shall consider in the next section and in Part II. But here we are interested in definitions of an orthodox sort; and when treated in this way, the definition is incapable, as it stands, of determining the domain M on the basis of the subdomain I of individuals. For it contains no predicate either for M or for I and hence nothing that could be taken to determine M or to be determined by I.

But this defect is readily remedied. For we may suppose that the variables for concepts and objects that appear in the statement of the Law are relativized to a predicate D for the domain. Thus instead of saying 'for all concepts', we say 'for all concepts defined over D'; and similarly for the other locutions. We may also introduce a predicate I for being an individual and add to Hume's Law the assertion that every individual is in the domain (i.e. every I is a D). With these changes, Hume's Law might not only be taken to determine a number operation F_0 on the given domain M_0 of all objects; it might also be taken to determine a domain $M = g(I)$ on the basis of the subdomain I and a number operation F_M on the basis of M. It can be treated, in other words, as a definition of F and M in terms of I.

So Hume's Law, and other such abstraction principles, can be taken to perform these two definitional tasks. But how well? To what extent are they of help in determining the domain of objects or the operation of abstraction?

The answer is 'very badly'. The abstraction principle, even when modified, is unable to determine a unique expansion M of any given subdomain of individuals I. For given that it has determined one expansion, it will be incapable of excluding any other expansion with the same number of abstracts (or abstract-surrogates).

The principle is also generally incapable of determining a unique abstraction operation on a given domain. For suppose that Φ is an abstraction principle concerning a certain abstraction operator §; and let us replace the operator § in Φ by an appropriate variable F to obtain the general condition Φ(F). Then Φ(F) is never uniquely satisfied over a given domain (except in the case in which the domain contains only one object). For let *F* be one solution to Φ(F). Let ∨ be the universal abstract, i.e., the abstract $F(\bigvee)$ where $\bigvee$ is the universal concept. Let *x* be any other object in the domain (possibly even an abstract). Permute ∨ and *x*. The resulting operation *F′* will then be distinct from *F* and yet still be a solution to Φ(F).[12]

In the light of the general importance of this phenomenon, it will be helpful to pinpoint more exactly how and where the indeterminacy arises. It can arise in two separate ways. One concerns existence. Is there, for any subdomain of individuals, an expansion of the subdomain with abstracts and an interpretation of the abstraction operator for which the abstraction principle is true? The answer, at least for the non-circular principles, is clear; there will exist such an expansion and interpretation as long as the the principle is non-inflationary over arbitrarily large domains.

Second, there is the question of uniqueness (aside from existence). Are there different ways of expanding a subdomain of individuals? And are there different interpretations of the abstraction operator that can be imposed over a given domain?

Uniqueness, in either case, can fail for two rather different reasons, one structural and the other 'material'. There can be structurally different, i.e. non-isomorphic, solutions. But even if all the solutions are structurally the same, there may still be solutions that differ in how the structure is realized. The one is a failure of categoricity, and the other a failure of determinacy within a given isomorphism-type.

Let us see how each source of failure might manifest itself in the case of abstraction principles. Consider the structural question first. In regard to the problem of domain expansion, the question is whether the number of abstracts is determined by the number of individuals. To this the answer is no. Hume's Principle, for example, is compatible, in the absence of any individuals, with there being a countable infinity of numbers or with there being an inaccessible number of numbers.

---

[12] The more usual method of proving non-uniqueness uses Frege's switching argument (cf. Lemma III.37). But the proof breaks down when the identity criterion is circular.

Let us suppose that the structural problem of domain expansion has been solved; it is known how many abstracts go with each predetermined number of individuals. Then in regard to the problem of interpreting the operator, the structural question that remains is whether, given the number of individuals and the number of abstracts, the interpretation is determined up to isomorphism.

The answer is again, no. Suppose, for example, that Φ is a principle that takes each finite concept into its extension and every other concept into the same abstract ('Infinity'). Then such a principle is compatible with the finite extensions being well founded over a given domain and also compatible with their not being well founded (see the comment to Theorem III.6.5). It turns out that the answer is yes for number abstraction and for the various weaker forms of abstraction (Theorem III.6.6); and this is perhaps the reason why this particular source of indeterminacy has been overlooked.

Consider now the question of identity. In regard to the problem of domain expansion, this is partly a question of ascertaining which objects are abstracts of the kind in question and which are individuals. We can then be sure that the expansion is always from individuals to abstracts of the right sort. But the question is also partly a matter of getting the abstracts right in any given case. For example, in the case of Hume's principle, we only want to include a number in the expanded domain as long as all smaller numbers are included.

A principle of abstraction is powerless to settle the question. For take any solution g to the problem of domain expansion. Then we may replace abstracts and individuals at will, always taking care to replace distinct objects with distinct objects, and we will still have a solution.

Let us now suppose that the identity problem for domain expansion has been solved; we know which abstracts go with which individuals. Then in regard to the problem of interpreting the operator, the identity question that remains is the question of determining which abstracts go with which concepts. Thus whereas the first identity question is a matter of getting the range of the abstraction operation right, the second is a matter of getting the values within the range right, of correctly assigning them to the concepts.

Again, a principle of abstraction is powerless to settle the question. For, as we have seen, we may permute the universal abstract with any other and still retain a satisfactory interpretation of the abstraction operator.

The general problem of identity corresponds to what has been called the 'Caesar problem'. This is the problem of determining the truth-values of sentences of the form 'the number of Fs = t', where t itself is not of the form 'the number of—' (and similarly for other forms of abstraction). The exact relationship between the two problems depends upon what terms are admitted as substituends for t. Suppose that each object in the domain is taken to have a name whose designation remains the same regardless of the interpretation of the abstraction operator. Then the two issues are essentially the same. For example, if '0' is a rigid designator (in this sense) of the number 0, then the truth of 'the number of non-self-identicals = 0' will guarantee that the correct object is assigned to the empty concept.

However, in discussions of Caesar's problem it is usually assumed, if only implicitly, that any designator for a number will be reducible to one of the form 'the number of—'. This assumption is, in fact, quite controversial. It requires a logicist account of arithmetical operations, for an arithmetical expression like '3 + 5' must be understood in some such way as 'the number of the disjunction of two exclusive concepts of which 3 and 5 are the respective numbers'; and in the case of an expression such as 'Caesar's favourite number', it becomes even more problematic whether a reduction to the required form can be given.

In any case, as long as this assumption is made, the two problems will be inequivalent. For the 'Caesar' term t must then always designate an individual; and so a solution to the Caesar problem will merely guarantee that the abstraction terms refer to abstracts, not that they refer to the right abstracts. We only have a solution to what might be called the 'external', as opposed to the 'internal', Caesar problem.

In the face of these various forms of indeterminacy, what kind of definitional status should we accord to Hume's Law and the various other abstraction principles? One possible response to the indeterminacy is to say that it is no worse than the indeterminacy that besets our actual conception of number or of some other kind of abstract. The definitions provided by the various abstraction principles may therefore reasonably be taken to be indeterministic in character. Thus Hume's Law may be taken to determine a variable domain of objects on the basis of the individuals, one that will allow different choices of objects for the numbers; and it may be taken to determine a variable number operation for a given domain, one that will allow different assignments of the objects from the domain to the concepts. It may then be maintained that our understanding of the numerical locutions, when so defined, exactly accords with our ordinary understanding of those locutions.

This is not the view I am myself inclined to take. For there seems to be a striking difference between our conception of numbers, on the one hand, and our conception of an axiomatically defined class of elements, such as a group, on the other. The indeterminacy that infects our conception of the unit element *e*, for example, seems to be of a quite different order than that, if any, which infects our conception of the number 1. Thus whereas it is acceptable to let *e* 'be' 1 under a suitable interpretation of the group operator, it makes no sense to suppose that 1 might be the undetermined group element *e* under a suitable interpretation of the numbering operator. Any view must therefore be judged unsatisfactory if it does not explain the asymmetry between the two cases.

It is not my aim to debate the merits of the indeterministic conception here.[13] I shall merely take for granted that some form of deterministic conception is correct. In this case, it must be conceded that the mere requirement of conformity to an abstraction principle is incapable of yielding our ordinary conception of the kind of abstract in question; and it is therefore incumbent upon the advocate of the abstractionist approach to provide an account of what else might be relevant in determining our understanding of the kind.

There are two main forms of response that might be pursued. One is to supplement the principle with further conditions. The principle would then be only part of the full definition; it would be the definitional analogue of an enthythematic argument. Under such an approach, it may be asked why the principle, rather than the supplementary material or the full condition, should be taken to define the operator in question. But, as we shall see, the principle may provide what is distinctive about the full defining condition and

---

[13] However, let me note that the indeterministic conception of number may be used to evade the dilemma that Crispin Wright (1988: 450–1; taken from his 1983: 148–52, n. 20) poses for Field. He asks: how is it possible to explain number using Hume's Law and yet deny the existence of numbers? To which the answer is: by treating Hume's Law as an explanation of a *variable* number operator. The existence of numbers may then intelligibly be denied since the denial simply amounts to the claim that there is no operator that conforms to the Law. If we regard Hume's Law as part of a 'scientific' theory, then this response is equivalent to a Ramsey-style treatment of the theoretical terms.

there may be a uniform method for going from the principle to what else is in the condition.

The other line of response is to strengthen the requirement of conformity. Instead of requiring mere conformity to the principle, we require conformity of a special sort. The principle alone then serves to define, but in a way that goes beyond its mere satisfaction.

The distinction between the two forms of strengthening can be illustrated with the case of inductive definition. Consider the inductive definition of *even* number: 0 is even; and for each natural number n, $n + 2$ is even if n is. Mere conformity to these rules is not sufficient to determine the required interpretation of 'even'. All the same, the rules do provide a basis for a satisfactory definition of the term in either of the indicated ways. For we can either stick to conformity and supplement the given rules with the rule that any even number belongs to all classes that are closed under the given rules; or we can stick to the given rules and constrain conformity by the requirement that the extension for 'even' should be minimal.

As this example shows, it may be possible to strengthen a given definition in either of the two ways. And, in general, either kind of strengthening can be traded in for the other: for conformity to additional conditions can be regarded as a special kind of conformity; and conformity to the given conditions in a special kind of way can be regarded as an additional condition. Usually, however, it will be more natural to regard a definition in one way rather than another.

How then might an abstraction principle be supplemented or the manner of conformity strengthened so as to yield a determinate conception of the domain of abstracts and of the abstraction operation?

A solution of sorts to the structural problem may be given in terms of the generative model of abstraction that is developed in sect. III.5. Suppose that we are given a domain of individuals of fixed cardinality. Then we may successively generate the abstracts over that domain in a natural manner. In the case of number abstraction, for example, we may first generate the numbers of the various concepts of individuals. We may then add these numbers to the given domain and generate the numbers of the various concepts over the augmented domain. Continuing in this way, we obtain a domain and a partial interpretation of number abstraction at each finite stage. By taking the limit of these partially interpreted domains, we may take the construction into the transfinite, stopping at the first stage at which no new numbers are introduced.

What such a construction rules out is the possibility of an 'ungrounded' abstract. Any number, for example, must be the number of some concept. Moreover, it must be possible so to choose the concept that it can be specified without reference, either direct or indirect, to the number in question. Continuing in this way, the identity of every number must ultimately be grounded in the identity of the individuals.

The construction will give us a categorical interpretation of the number operator over the given domain; any two interpretations that are generated in this way will be isomorphic as long as the cardinalities of their respective domains of individuals are the same. Moreover, the result will generalize. The condition of being in one–one correspondence that figures as the identity criterion in Hume's principle has the property of being absolute: whether two concepts are related by the condition depends only upon the objects that fall under the concepts in question, not on those that fall outside the concepts. The result will also hold for any abstraction principle whose identity criterion is absolute in this sense.[14]

This solution may be characterized under each of the two heads for completing a definition. On the one hand, we may suppose that Hume's principle, to continue with the example, is supplemented by the principle that there is no proper subdomain that contains all individuals and contains the number of any concept of objects in the subdomain. Or we may require that the solution be minimal in the sense of not containing a smaller solution. Thus the way the completion goes is not essentially different in this case than in the case of inductive definition.

The above generative model can be extended to apply simultaneously to several abstraction operators. We might imagine, for example, that numbers and finite extensions are generated in tandem. But, more significantly, the means of abstraction can

---

[14] Our account here is reminiscent of the generative model of understanding that was originally proposed by Wright (1983: 142–5) and subsequently elaborated by Wright (1998a) and Hale (1994) in response to criticisms of Dummett (1991a: 236). Wright's account is not intended as a way of generating the domain of all numbers but as a way of showing how we can understand terms for all the natural numbers. It is further discussed in sect. II. 6. Wright (1997) gestures towards the notion of absoluteness at the close of sect. II, but in regard to the question of acceptability rather than of impredicativity or categoricity.

themselves be taken to be subject to generation. Let us suppose that we start off with a given domain. We may then generate the acceptable means of abstraction over that domain, i.e. the ones that are non-inflationary and predominantly logical. We may then apply these means of abstraction to the concepts over the domain to obtain new abstracts. These abstracts are then added to the domain and the whole process repeated ad infinitum.

Such a construction yields not only a determinate conception of what abstracts there are, given the means of abstraction, but also a determinate conception of what means of abstraction there are. It therefore solves the problem of determining what the cardinality of the universe of abstracts should be; for the cardinality is determined by the point at which the process yields nothing new (sect. IV. 2).

The above solution is very natural and, indeed, it is hard to think of reasonable alternatives. Thus as long as the appropriate minimality assumptions are made, Hume's Law—and other such principles— will provide us with a complete structural understanding of the abstracts with which they deal. The problem of identity, however, is not so tractable. Even if it is agreed that the abstracts are generated from the individuals in the manner proposed, we still need to say what the abstracts are. One solution to this problem is implicit in the general theory of abstraction that is developed in Part IV. The basic notion of this theory is that of an object being the abstract *of* a concept *with respect to* a relation on concepts. Thus the notion relates three items: an object, which is the result of the abstraction; a first-order concept, which is what is abstracted; and a second-order relation on concepts, which is the means of abstraction. In the case of numbers, for example, we may say that 0 is the abstract of the empty concept with respect to the relation of one–one correspondence.

In order to solve the identity problem, we can simply declare that the number of *C*s is the abstract of *C* with respect to the relation of equinumerosity, that the extension of *C* is the abstract of *C* with respect to coextensiveness, and similarly for the other cases of abstraction. Such an account provides us with an explanation of the particular kinds of abstract in terms of the general notion of an abstract; and, to the extent that the identity of the general abstracts is clear, then so is the identity of the particular kinds of abstract.

This solution can be regarded as a generalization of Frege's account of the number of *C*s as the extension of the concept of being equinumerous with *C*. For upon treating the abstract of a concept with respect to a given equivalence relation, in our account, as the class of all concepts equivalent to the given concept, we obtain the Fregean abstract. However, our viewpoint is quite different from Frege's. For he took there to be a privileged form of abstract, one to which all others were reducible, whereas we take the different forms of abstract to be *sui generis.*

Indeed, from the standpoint of a general theory of abstracts, the Fregean view is incoherent. For he identifies each conceptual abstract, in respect to a given means of abstraction, with the corresponding extension or class of concepts. But the extensions are themselves among the abstracts. So, by the same general rule, each extension should be identified with the class of all the concepts that have that extension, which is absurd.

We have obtained solutions to the problems of identity and of structure. However, the two solutions do not sit well together. For given our solution to the problem of identity, we already have a complete explicit definition of number as a certain sort of abstract; and in the presence of an explicit definition of an operator it is hard to see what further role might be served by an implicit definition. Our definition of number in terms of abstracts, like Frege's in terms of extensions, should be regarded as an alternative to the definition in terms of Hume's Law, not as an addition to it.

This is not to deny the analyticity of Hume's Law, or of whatever other kind of abstraction principle might be in question. For even when the principle does not follow directly from the explicit definition of the abstraction operator, it may still follow from the definition and the analytic principles governing the notions in terms of which it has been defined. (Indeed, this is just what Frege hoped for in the case of Hume's Law). Thus an abstraction principle may still be indirectly definitive of its operator even though it forms no part of its explicit definition.

But is there any reasonable way in which Hume's Law could be taken to contribute directly, as part of an explicit definition, to our understanding of the number operator? One possibility is to appeal to the essential properties of numbering and of the numbers. Suppose we call an operation 'quasi-numerical' if it conforms to Hume's Law. As we have observed, many different operations are quasi-numerical. By adding further conditions to Hume's Law, we want to distinguish the (genuinely) numerical operation, the one that takes each concept into its number, from the others. Once we have successfully defined what it is to be a numerical operation, we can take the number of *C*s to be the result of applying this operation to *C*.

Now one way the genuinely numerical operation is distinguished from some of the others is that it is *essentially* quasi-numerical; it is, by its nature, quasi-numerical. Consider, for example, an operation of being a 'qumber', defined by the condition:

the qumber of Fs is the number of Fs if that number is not 9 and otherwise is the number of the planets.

Then although the qumber operation conforms to Hume's Law it only does so accidentally.

This condition is not enough, on its own, to distinguish the genuinely numerical operator from all the others. To see why, define the operation of being a schumber by the condition:

the schumber of Fs is 0 if F is singleton, it is 1 if F is empty, and it is the number of Fs otherwise.

Thus schumber is like number, but with the roles of 0 and 1 reversed. It may then plausibly be maintained that the operation of schumbering is also essentially quasi-numerical.

However, the genuinely numerical operation has a further distinguishing feature. For each value of the operator, i.e. each number, is essentially the value of the operation for suitable choices of the concept. Thus the number 0 is essentially the number of the concept of being non-self-identical, the number 1 is essentially the number of the concept of being identical to 0, and similarly for the other cases. This further feature distinguishes numbering from schumbering. For 0 is not essentially the schumber of any singleton concept; it is no part of the account of what 0 is that it should be related in any way to a singleton concept.

Moreover, this feature is itself essential to numbering. I therefore propose to define a genuine numerical operation as one that essentially conforms to Hume's Law and is essentially definitive of its values in the way described.

It is of the utmost importance, in evaluating this proposal, to distinguish between the definition of a linguistic expression and a 'real' definition of the non-linguistic item for which it stands. Thus the expression 'the number of objects that are not self-identical' may be taken either as a definition of the numeral '0' or as a real definition of the number 0. In the first case, the identity '0 = the number of objects that are not self-identical' is taken to be true because of the meaning of the numeral '0'; and in the second case, it is taken to be true because of the nature or essence of the number 0.

What we are proposing is that each number (not numeral) has a definition of the form 'the number of *C*s' (it is essentially of that form for suitable *C*). We are also proposing that the number operation (not operator) has a definition part of whose content includes the definability of the numbers. The predicate (not property) 'is a numerical operator' is then defined by reference to these non-linguistic forms of definition.[15]

I am not sure that the present definition of number in terms of its essential properties should be regarded as being at odds with the previous definition of number as a certain kind of abstract. It might even be regarded as providing a deeper account of the notion of abstract. For we may ask: What is it for *x* to be an abstract of a concept *C* with respect to a means of abstraction *R*? Our essentialist form of definition then provides an answer: for an abstraction operation can be singled out in terms of its essentialist relationship to *R* and the abstract *x* can then be taken to be the value of this operation for the concept *C* as value.

I turn finally to the question of how a definition of number in terms of Hume's Law might be of significance for the philosophy of mathematics. There are two major problems in the area that such a definition might be thought to be of help in solving. The first is semantical: it is to show how we can make determinate reference to numbers. The second is epistemological: it is to show how we can have knowledge of number-theoretic truths.

Consider the semantical problem first. In order for a definition of number to resolve philosophical doubts over determinacy, it must

---

[15] Let me make a few further comments on this proposal (a full discussion is out of place). (1) I do not presuppose an account of essence in modal terms and, indeed, am inclined to reject such an account for the reasons stated in Fine (1994). (2) For the purposes of the account, operations are best taken in an intensional sense though concepts may be taken in either an extensional or intensional sense. (3) There are other views as to what the definability of the numbers might consist in. One might maintain, for example, that the number 0 is essentially the number of any empty concept, that the number 1 is essentially the number of any singleton concept, and similarly for the other cases. Another possibility is that for *any* singleton concept, except one whose object presupposes the number 1 itself, the number 1 may be defined as the number of that concept. Of the three possibilities it is the third that is the most plausible. However, it involves a distinction between definitive and essential properties that is explained in Fine (1995) but which I have here preferred to ignore.

satisfy three desiderata. First, it must satisfy the formal requirement of having a unique solution. Second, the terms by which the number locutions are defined must not themselves suffer from a similar indeterminacy. Third, the definition must be correct and, in particular, must not rest upon an arbitrary identification of numbers with one kind of object as opposed to another.

Frege's definition of number as an extension of concepts is based upon an incoherent conception of extension and fails, in any case, to satisfy the third desideratum. Our alternative definition of number in terms of the general notion of abstract satisfies the first desideratum and arguably satisfies the third. For there is nothing arbitrary in the identification of numbers with a certain sort of abstract. However, the definition fails to satisfy the second desideratum. For the notion of an abstract is subject to the same sort of indeterminacy as the notion of a number. The problem of determinacy is merely pushed back.

Hume's Law, on its own, fails to satisfy the first desideratum; for, as we have seen, many different operators will conform to the Law. It must therefore be supplemented in some way if it is to be of any use. We have suggested a supplementation in terms of the essentialist role of the Law; and it is arguable that the resulting definition satisfies all three desiderata. It satisfies the second; for even though the notion of essence may be subject to indeterminancy, it is not clear how it would result in an indeterminacy in what the numbers may correctly be taken to be. It satisfies the critical part of the third; for the definition does not even attempt to identify number with some broader category of object. Therefore the only room for doubt lies in the essentialist intuitions upon which the correctness and uniqueness of the definition rest. One might think, for example, that 0 can with equal legitimacy be defined as the number of an empty concept or as the schumber of a singleton concept. But in the absence of any such doubt, it is hard to see how determinacy could reasonably be denied.[16]

Consider now the epistemological problem: how can arithmetical truths be known? One natural approach to the problem is to provide an explicit definition of the number operator in terms of other notions which, in the special case of logicism, will be logical notions. Arithmetic can then be derived from Hume's Law; and Hume's Law

---

[16] Even though the definition has these many virtues, I doubt that it can be regarded as ultimately satisfactory, since there should be a non-essentialist underpinning for the essentialist attributions upon which its correctness depends.

can be derived from principles governing the notions in terms of which number is defined. The problem of our knowledge of arithmetic therefore reduces to the problem of our knowledge of logic and of those other principles. When the principles are themselves logical, we have the familiar form of logicism and arithmetic is reduced to logic.

The possibility of implicitly defining number by means of Hume's Law might appear to give us an epistemological edge over the more usual forms of logicism. For Hume's Law is then secured by definition; and so we only need the logic to take us from the Law to arithmetic. But this is too simple. For whereas an explicit definition stands in no need of justification, an implicit definition does. An epistemological cost must therefore still be borne; but it lies now in the justification of the definition rather than in the justification of the Law on the basis of the definition.

Since the point is of some importance, it will be worth a more thorough discussion. To this end, let us ignore the question of whether the proposed definition is faithful to our ordinary conception of number and whether, in particular, it is referentially determinate. Our aim may be only to provide what might be called *reinterpretative* foundations; we aim to provide an interpretation of number that possibly deviates from the usual interpretation but will serve the same purposes and, in particular, will provide a basis for our knowledge of what goes for the usual arithmetical principles. Thus for these purposes, Hume's Law itself can be adopted, without supplementation, as an indeterministic definition of the number operator.

Now in order to derive arithmetic from such a definition, we must not only be able to endorse Hume's Law as a definition of number, we must also be able to make the transition from the endorsement of the Law as a definition to its endorsement as a truth. The interpretation assigned to the number operator by means of the definition must be such as to render the Law true.

Call a definition *materially effective* when this transition is warranted, i.e. when the making of the definition will result in its defining condition being true. Not all definitions are materially effective even when they are successful. For in case the definition has only semantical import, an interpretation will be assigned regardless of whether or not the defining condition is thereby rendered true. Should the definition be deterministic, the sense assigned may be that of an empty description; and should the definition be indeterministic, the sense assigned may be that of an empty variable. In neither case is it plausible to suppose that the resulting interpreted condition will be true.

There are, of course, various methods for evaluating a statement that contains an empty description or an empty variable; and some may result in such statements being true. For example, under the generality interpretation of the variable (according to which a statement containing a variable is true if it is true for all its values), all such statements will be vacuously true. But such a possibility is of no help in the present case. For we require that Hume's Law should be true under a conception of truth that more or less agrees with our ordinary conception of what is true; it should not render all arithmetical statements true or, in some other radical manner, cut across the usual distinction between what is true and what is false.

Thus if we initially conceive of Hume's Law as a definition, then in order to be justified in using it as a premiss in a demonstration of arithmetic, we must know that it is materially effective, i.e. that it results in a truth (in accord with our ordinary conception of what is true). But how can we know this? There appears to be only one answer. To know that the definition is materially effective we must know that it is referentially effective, i.e. that it succeeds in assigning a referent to the number operator in conformity with the Law (or its supplementation). But how can we know this further fact? Again, there appears to be only one answer. To know that the definition is referentially successful we must know that there exists an item that plays the role required of a referent. What this means, in case we take the definition to be deterministic, is that we must know that there exists a unique solution to Hume's Law (or its supplementation); and what it means, in case we take the definition to be indeterministic, is that we must know that there exists at least one solution to Hume's Law.

The use of Hume's Law as a definition is therefore not without epistemological cost; and whether it can secure an epistemological advantage and, in particular, whether it can be used to vindicate a form of logicism, is far from clear. For the only obvious way of justifying the existential generalization of Hume's Law (with respect to the number operation) is by means of an instance. We see that the generalization is true by taking the operation in question to be the number operation or some other, specific, mathematically defined operation. Thus even though the existential generalization is stated in purely logical terms, it is not clear that it has a purely logical justification or one that is in some other way less problematic than that of the Law itself.

3. Reconceptualization

I want to consider a non-classical approach to the question of defining number by means of Hume's Law. It might be called *definition by reconceptualization* and rests on the idea that new senses may emerge from the reanalysis of a given sense. The idea derives from sects. 63–4 of the *Grundlagen*. However, it has not been my aim to be faithful to Frege's thought; I have been content, in this matter, to follow the exposition of others, principally Dummett (1991*a*) and Wright (1983). My interest, rather, has been to see whether the ideas themselves can be sustained.

There is yet another non-classical approach to using Hume's Law as a definition of number. This is the context principle; and it rests on the idea that certain truths may be used to fix the reference of the terms that they contain. Thus whereas the first-mentioned approach rests on the adoption of an unorthodox mechanism for the determination of sense, the second rests on an unorthodox mechanism for the determination of reference. The second approach will be taken up in the second Part.

It is worth noting, however, that it is highly implausible that we might altogether avoid the epistemic cost involved in adopting Hume's Law as a classical implicit definition of number by adopting one of the less orthodox forms of definition in its place. For whatever the mechanism by which the sense or reference of the defined terms is secured, we surely require assurance that the form of words by which the definition is given can be asserted without danger of contradiction. This problem is especially acute in the light of the fact that a very similar form of words, that of Law V, does give rise to contradiction. Surely we require assurance that the two Laws are not alike in this respect if we are to have confidence in the conclusions obtained from the one as opposed to the other.[17]

---

[17] It is for this reason that I do not think that we can simply pass from the stipulation of Hume's Law as a definition to its assertion as a truth, without the need for further justification. Hale and Wright (2000b: sect.4) recognize the need for the definition to be consistent if the transition is to be safe, but fail to acknowledge that the definition-monger is himself under an obligation to show that the transition is safe.

But how is such assurance to be obtained? A *post facto* justification, one that already rests upon the acceptance of the definition in question, is useless; for an inconsistent definition is as capable of justifying its own consistency as a consistent definition. We could, of course, appeal to the corresponding existential claim (that there exists an operation in conformity with the Law). But, as we have seen, it is quite unclear whether this provides us with any epistemic advantage; and it would not, in any case, provide us with any advantage over the adoption of a straight implicit definition.

If we really are to provide a distinctive foundation for arithmetic on the basis of one of the novel forms of definition, then some mark of consistency must be stated and justified without implicit appeal to the abstract objects in question. We are therefore in much the same position as a nominalist who is required to defend the consistency of a given branch of mathematics. *Perhaps* this can be done. After all, one might follow the lead of Field (1989) by taking consistency to be a modal primitive and establishing the claim of consistency by some form of non-mathematical induction. Also, procedural postulationism (which I have not discussed) is able to provide a deductive basis for various consistency claims. However, there is nothing in the two non-orthodox forms of definition that we shall consider that provides any clue as to how consistency might be established or how it might serve to warrant the transition from the stipulation of a definition to the assertion of its truth.

Let us now turn to the question of whether number can be defined through reconceptualization. The basic idea behind this approach is that we may understand the left-hand side of Hume's Law as another way of saying what is said on the right and in such a way that the number operator is thereby endowed with sense.

In discussing this approach, it will be important to distinguish between a universal and a schematic formulation of Hume's Law. A universal formulation is given by a closed sentence:

For all concepts F and G, the number of Fs = the number of Gs iff the Fs and the Gs are equinumerous.

The schematic formulation is given by a scheme:

the number of Fs = the number of Gs iff the Fs and the Gs are equinumerous, where 'F' and 'G' are now schematic letters for predicates, not variables for concepts.

What is meant by talk of the 'left' and 'right' sides of Hume's Law? Under the schematic formulation, the left side is presumably a (closed) identity statement of the form 'the number of Fs = the number of Gs'; and similarly for the right. But under the universal formulation, there is no *sentence* on the right or left. Presumably, then, the left-hand side is the *open* sentence 'the number of Fs = the number of Gs'; and similarly for the right.

How should reconceptualization be understood—as relating to the schematic or to the universal formulation? It might be thought not to make much difference, since the opportunities for reconceptualizing what is expressed on the right are much the same whether we have a closed or an open sentence. But this is not in general correct. Consider Frege's paradigm of objectual abstraction:

the direction of L = the direction of M iff L and M are parallel.

If the lines here are the lines of abstract Euclidean space, it is not clear that any reference to any particular one of them is possible. (It is idle to fix an origin and axes, since that already requires that we be able to single out a point and some lines.[18]) But if that is so, then there are no relevant closed sentences whose senses can be reconceptualized. It is therefore the open sentences that should be taken to be the target of reconceptualization.

Usually, when philosophers have talked of reconceptualization they have supposed that the sense of the right- and left-hand sides is a thought, i.e. the sense of a sentence. But if I am right, it would be better to take the sense to be an incomplete thought, i.e. the sense of an open sentence or formula.

We may divide the process of conferring a sense on the number operator into two stages: first, the formula on the left is taken to have the same sense as the formula on the right; second, this sense is subject to reanalysis in such a way as to provide a sense for the number operator. In a way, the method is a combination of the methods of explicit and implicit definition. For if the structure of the left-hand side is ignored (apart from the presence of the variables

---

[18] It might be doubted that there is even *in principle* a way of referring to one of the lines, or one of the points, as opposed to the others. I note, by the way, that this raises a serious problem for the usual development of Fregean intensional logic. For it means that we cannot assume that for each object there is a sense that picks it out.

'F' and 'G'), we may use the method of explicit definition to endow it with the same sense as the formula on the right. On the other hand, if we do not require the synonymy of the left and right sides, we can use the method of implicit definition to confer a sense on the number operation. The idea behind reconceptualization is to get the implicitly defined sense, not directly from the principle, but indirectly from the explicitly defined sense.

Does the method work? Various questions should be distinguished here. Can we obtain a unique sense for the number operator in this way? Can we obtain at least one sense, even if not a unique sense? And, in each case, what is required by way of justifying the definition?

It should be noted that if the definition succeeds in conferring a unique sense on the number operator then the definition is referentially effective and, indeed, referentially determinate. For if the sense did not correspond to a referent or if the referent were an operation on concepts that were not always defined, the formula 'the number of Fs = the number of Gs' would presumably lack a truth-value for certain concepts as values for 'F' and 'G' and so its sense could not be the same as that of 'the Fs are equinumerous with the Gs'. Thus the unique sense delivers a unique referent; and there seems to be nothing in the sense that would allow for any indeterminacy in the referent that was so determined.

This point is of some importance, because it means that the method is more powerful than it is commonly taken to be. Several authors, I suspect, have thought of definition by reconceptualization and the context principle as linked in the defence of logicism. Perhaps the thought is that the one provides a route to sense and the other a route to reference. But since reconceptualization already yields reference, it can be allowed to stand on its own.

A related point holds in regard to sects. 64–6 of the *Grundlagen*. Frege there considers three objections to the proposed method of defining the number operator. The first two he dismisses, but the third he upholds. It is that the definition is unable to settle the question of whether Caesar is a number. But his objection appears to be misguided. For granted that the sense of the number operator has been fixed, that the recarving of content has been successful, the operation that is the referent of the operator will have been fixed; and so it is determined whether or not Caesar is a value of this operation. Of course, at this point Frege did not have the distinction between sense and reference; but the point remains that the 'concept' of the number operation can hardly have been fixed without its values also having been fixed (cf. Frege's remark 'what we lack is the concept of direction'). It is possible that in posing this objection Frege is here reverting to the model of implicit definition; for an implicit definition of number by Hume's Law will, of course, fail to settle the question of whether Caesar is a number. Alternatively, it is possible that objection is meant to show that the proposed recarving will not work. Since it is evident that the Caesar question is not resolved, there must be something wrong with the proposed attempt to recarve, even if it is not clear what it is.[19]

Does definition by reconceptualization, then, deliver a unique sense? I am not sure that there is anything in principle to be said against the possibility of such a definition fixing sense. To take an example of Dummett's (1991*a*: 173), suppose we attempt to define the predicate 'commits suicide' by means of the condition:

$\forall x$ (x commits suicide iff x kills x).

Then it might reasonably be thought that the definition works by reconceptualization: there is a sense expressed on the right, which is the result of linking the argument-places in the sense of the predicate 'kills'; and this sense is then reconceptualized as that of a one-place predicate, with a single argument-place rather than two linked argument-places.[20]

There are, however, reasons of detail against supposing that anything analogous could work in the case of Hume's Law. Consider Law V, its inconsistency being irrelevant for present purposes. Suppose the Law is given in the formulation $\S x Cx = \S x Dx \leftrightarrow \forall x \ (Cx \leftrightarrow Dx)$; and let us adopt the following abbreviations:

[19] This is how Dummett (1991*a*: 126) appears to construe him although on pp. 168–9 he also takes him to be making a claim of synonymy. I might note that if Frege had an implicit definition in mind, he might also have objected that it fails to serve his epistemological purposes, for its existential presupposition would stand in need of justification.

[20] Cf. Frege's *Begriffsschrift*, sect. 9 and the discussion in Dummett (1991*a*: 73–5 and 1981: 333) and Hale (1997: sect. 3). Dummett's (1991*a*: 174–5) account of the way the definition might work in this case is somewhat different from my own. He supposes that there is a 'linguistic device' that converts the LHS into the RHS and two corresponding notions of sense, one coarse and the other more refined. The coarse sense is only assigned to an expression after the linguistic device has been applied; and so the coarse sense of the two sides is the same. The refined sense of the LHS also includes 'the functioning of the device'; and so the refined senses of the two sides will be different. Thus he does not posit a single sense that is subject to two different analyses.

(A) for ∀x (Fx ↔ Gx);
(B) for §xFx = §xGx;
(C) for ∀x (x = §xFx ↔ x = §xGx); and
(D) for §x (x = §xFx) = §x (x = §xGx).

Then on the view under consideration, (B) should have the same sense as (A) and (D) the same sense as (C). Now presumably (C) should also have the same sense as (B). Indeed, it is hard to see how a criterion of synonymy could be so tolerant as to allow the synonmy of (B) to (A) (or of (D) to (C)) and yet not tolerant enough to forbid the synonymy of (C) to (B). If this is right, it follows that (A) has the same sense as (D). But then if the sense of the operator given through reconceptualization is to be unique, the sense of §xFx should be the same as the sense of §x (x = §xFx), which is presumably not so.

A similar, though more complicated, line of argument can be constructed for the case of number. Let F^+ be the concept of falling under F or of being the least natural number not to fall under F; and similarly for G^+. Let (A), (B), (C), and (D) now be as follows:

(A) for 'the Fs are equinumerous with the Gs';
(B) for 'the number of Fs = the number of Gs';
(C) for 'the F^+s are equinumerous with the G^+s';
(D) for 'the number of F^+s = the number of G^+s'.

It is then plausible to suppose, especially given the synonymy of (A) to (B), that (A) is synonymous with (C). But from this it follows that 'the number of Fs' and 'the number of F^+s' will be synonymous, which is impossible since, for any finite concept F, the number of F^+s will be the successor of the number of Fs.

In the light of such examples, it is difficult to see what general principles governing the identity of sense might permit its reconceptualization and yet prevent its proliferation. Thus it seems to me that the attempts of Wright (1988: 459) and Hale (1994; 1997) to identify a suitably weak sense of sameness of sense that might admit of reconceptualization are doomed to failure.[21]

But let us ask whether there exists at least one sense of the number operator for which the two sides of the Law will be synonymous or otherwise equivalent. In this case, it will certainly help to weaken the

---

[21] Indeed, if one examines the details of Hale's proposal towards the end of the postscript to chap. 4 of Hale and Wright (2000a), one sees that it allows proliferation, since the equivalences between (A) and (C) in each of the two cases is 'compact'.

relationship. For suppose we adopt an extensional conception of sense, under which sense and reference are to be identified. Then any of the many operations for which Hume's Law holds will provide a 'sense' for which the two sides are 'synonyms'. It might even be possible to maintain the existence of a sense under the more reasonable construal of synonymy as analytic equivalence. For if logicism is true, it will provide a sense of the number operator for which Hume's Law is analytic; and, for this sense, the two sides of the Law will, in the required way, be synonyms.

But the difficulty now concerns not the actual existence of the sense, but the grounds that we have for supposing it to exist. It is worth reminding ourselves that Law V is inconsistent. Thus it cannot be on the basis of any simple formal considerations that the existence of a sense is to be ascertained. But then what grounds can there be for supposing the required sense to exist? One should not be dogmatic on such a matter, but it is hard to believe that the existence of a sense for which the two sides are, in some strict sense, equivalent could be any less problematic than the existence of an operation for which the two sides are materially equivalent. Thus it seems that the possibility of definition by reconceptualization can take us no further, in this regard, than an implicit definition of a standard sort.

4. Foundations

Frege attempted to provide a logical basis for arithmetic and analysis. All the notions of arithmetic and analysis were to be defined in logical terms; and all the theorems of arithmetic and analysis were to be derived, with the help of the definitions, from logical axioms. The attempt failed if for no other reason than the inconsistency of the critical axiom, Law V.

Recently, Crispin Wright (1983) has attempted to revive the Fregean logicist programme. He observes that the Fregean derivation of arithmetic may be broken down into two steps: first, Hume's Law may be derived from Law V; and then, arithmetic may be derived from Hume's Law without any help from Law V.[22] He therefore suggests that Hume's Law be used instead of Law V as a logical basis for arithmetic.

[22] For details see Heck (1993), who suggests that Frege himself was probably aware of the possibility of separating the two steps. The possibility of such a derivation was first explicitly asserted by C. Parsons (1964: 194).

One difficulty with this approach is that it is unclear, in the absence of Law V, how to obtain a logical foundation for (higher-order) analysis. We can no longer treat reals as extensions of number concepts or the like; and nor is it clear what other form of abstract should be used in place of extensions. One might try experimenting with restricted forms of extensionality; but they appear either to be inconsistent or too weak or ad hoc.

Perhaps the closest we come in the existing literature to a satisfactory formulation is with Boolos (1986–7; 1993). He proposes (1986–7) that extensional abstraction be permitted on *small* concepts, i.e. on those that cannot be put into one–one correspondence with the universal concept; and he extends the proposal to those concepts whose complements are small even when they themselves are not (1993). He shows how the natural numbers can be identified with representative finite extensions, such as those considered by Zermelo or von Neumann. However, he cannot adopt Frege's definition of number as the extension of a second-level concept; and nor is it clear that he can provide any other definition of number that is capable of yielding Hume's Law in its full generality. The system is in a certain respect too weak; since the reals can only be obtained by adding some ad hoc axiom to the system, such as that the concept of being a natural number is small. And the system is in another respect too strong; since it only allows a standard model of cardinality c when c is an almost strong regular limit cardinal or a successor cardinal d^+ for which $2^d = d^+$.[23]

Of course, the reals might simply be 'posited' as the limits of rationals. Thus we might add to the language a limit-operator on concepts and a restricted criterion of identity to the effect that, when the concepts C and D of rationals have rational upper bounds, the limit of the C's is to be identical to the limit of the D's just in case the rational upper bounds of the one are the same as the rational upper bounds of the other. However, such a principle seems entirely ad hoc; and it would be desirable to have a more general basis upon which it might be established.[24]

In Part IV, I propose what I hope is a natural solution to these problems. I first present a general theory of abstraction, one which is

---

[23] The cardinal c is *almost* strong if $2^d \leq c$ whenever $d < c$. See Shapiro and Weir (1999). (I mistakenly said 'strong' in the original paper.)

[24] This approach has recently been pursued by Hale (2000), who couples it with a Frege-style account of reals as ratios of quantities. He attempts in sect. 4 to reduce the element of

independently plausible and which, in fact, was developed without regard to foundational considerations. I then show that higher-order analysis can be developed within the theory under a suitable identification of reals with abstracts.

The reals are not identified with extensions but with what one might call *generalized numbers*. With each concept of numbers, one may associate a criterion of identity, one by which two concepts F and G are related when the number of Fs and the number of Gs both fall under the given concept or both fall outside the given concept. The abstracts corresponding to these criteria of identity can be shown to exist within the theory; and it is these abstracts, rather than the corresponding sets of numbers, that do duty for the reals.

A distinctive feature of the theory is its generality. Instead of attempting to account for this or that form of abstraction, we attempt to account for them all. Thus the basic notion of the theory is that of an object being an abstract of a concept with respect to a given possible means of abstraction; and the aim of the theory is provide an axiomatic characterization of all those possible means of abstraction that actually give rise to abstracts.

There are two basic postulates of the theory: one concerning existence and the other concerning identity. The existence postulate says, roughly, that as long as the means of abstraction is non-inflationary and predominantly logical it will give rise to abstracts; and the identity postulate, in its strong form, says that two abstracts are the same if and only if they are associated, through their respective means of abstraction, with the same equivalence classes of concepts. For the purposes of foundations, however, the two postulates can be weakened: the existence axiom can be confined to the logical criteria; and the identity can be restricted to its left-to-right direction.

The question of the identity of abstracts will be discussed in the next section; and the question of their existence has already been taken up. But we should note, in connection with our foundational aims, that the instances of the existence axiom are conditional in

---

adhocery but, as far as I can see, does not come up with a fully worked out proposal. If Cut Abstraction (the limit principle) is restricted to a small domain of underlying objects, then we must assume, or somehow show, that the domain of rationals is small; and if it is restricted to domains that can be specified by means of a sortal concept (an idea also canvassed by Wright (1999: sect.3), then we shall need axioms that enable us to establish the existence of a suitably wide range of sortal concepts (the notion of sortal concept in the resulting theory would perhaps play a similar role to the notion of set in NBG).

form. An instance of an abstraction principle, for a given method of abstraction, is only asserted under the condition that the method is not inflationary, i.e. does not result in more abstracts than objects. Thus Hume's Law cannot be asserted in its categorical form but only under the condition that the universe is already infinite.

This means that we must reverse the procedure adopted by Frege. Instead of using Hume's Law to show that the universe is infinite, we must use the fact that the universe is infinite to establish Hume's Law.

But how can this be done, given the absence of any strong categorical forms of abstraction? The answer lies in the generality of the theory and in the additional assumption, which we make, of the existence of at least two objects. For we can then prove the legitimacy of an infinite number of means of abstraction, each of which produces at least two abstracts. By the identity postulate, it can then be shown that all these different abstracts are distinct. Thus it is from the diversity in the allowable means of abstraction, rather than from the power of any one of them, that the theory derives its strength.

If the diversity of means is genuinely to yield a diversity in abstracts, however, it is essential that abstracts that derive from different means of abstraction should not be indiscriminately identified. One and the same abstract is not to wear different 'faces' in connection with different means of abstraction. Thus a satisfactory solution to the Caesar problem is essential to our foundational programme and is not, as it is for others, something required merely for philosophical purposes.

There is another approach to the question of foundations that is suggested by our work and is more in line with modern thinking on the subject. For we can treat the objects of our theory as classes rather than as abstracts and dispense with any reference to an underlying criterion of identity. Thus we no longer regard an 'abstract' as being associated with various items by means of a method of abstraction but as simply being a collection of those items. On this view, although every abstract is taken to be a class, not every class is taken to be an abstract with respect to the coextensionality of concepts. Thus our previous difficulty over identifying abstracts with classes does not arise.

There are somewhat different ways in which this alternative approach might be presented. Perhaps the most orthodox, by current standards, is to take the notion of membership as primitive. The setting for the theory is then second-order logic, and comprehension for classes of concepts takes the form:

$\exists y \forall F$ (F ϵ y $\leftrightarrow$ ϕ(F)), where ϕ(F) is predominantly invariant.

(The condition that ϕ (F) be predominantly invariant is something that can be expressed within the theory itself). One can formulate a similar comprehension principle for classes of objects:

$\exists y \forall x$ (x ϵ y $\leftrightarrow$ ϕ(x)), where ϕ(x) is predominantly invariant.

However, in order to obtain a theory of reasonable strength, it is essential to countenance the application of the principle to classes of concepts and not just to classes of objects.

One might see these principles as arising from a certain natural approach to the problem of impredicativity. Consider the universal class V. One may define it as the class of all objects x for which $x = x$. Now there is a sense in which this definition is impredicative, for the universal class V is already included within the range of the variable x. But there is also a sense in which it is not. For one can understand what the condition is without presupposing any knowledge of the class V. Indeed, our understanding of the condition $x = x$ is not something that requires the understanding of any particular objects at all. Once we have defined the class V, we may then define other classes with its help, such as the unit class of V, or the class of all objects not identical to V; and so on. In general, it seems reasonable to suppose that we can define any class by means of a condition as long as our understanding of it does not involve too many objects. We can think of our comprehension principle as a way of making this view precise: the idea of a condition involving an object is explained in terms of invariance; and the idea of 'too many' objects is explained in terms of an appropriate notion of smallness.

An alternative method of presentation, more in keeping with Frege's own, is to take an extension-forming operator § as primitive and then to adopt the following restricted form of abstraction for extensions:

§Fϕ(F) = §Fψ(F) $\leftrightarrow$ $\forall$F (ϕ(F) $\leftrightarrow$ ψ(F)), where ϕ(F) and ψ(F) are both predominantly invariant.

Similarly, in the case of objectual extensions, the principle would take the form:

$\S x\varphi(x) = \S x\psi(x) \leftrightarrow \forall x \, (\varphi(x) \leftrightarrow \psi(x))$, where $\varphi(x)$ and $\psi(x)$ are predominantly invariant.

Within such a theory, under either presentation, one can then provide a foundation for both arithmetic and analysis. Hume's Law can be derived in much the same way Frege derived it from Law V, using the very same definition of the number operator; and analysis can then be derived along the same lines as for the general theory of abstracts.

I do not regard these theories as providing support for Frege's overall logicist aims. If one adheres to the Context Principle, then one can perhaps conceive of the various abstraction principles as providing a definition of the notion of natural number and perhaps also of the notion of real number. Moreover, if one takes the notions of second-order logic to be logical, then these definitions will be stated entirely in logical terms; and this makes it very reasonable to suppose that the defined notions will then be logical themselves. But the logical status of the notions will not provide the defining principles with any special epistemological status; for, even though they are logical definitions, they will still stand in need of an extra-logical justification.

Perhaps the best that can be said on their behalf is that they provide us with some sort of theoretical insight into the nature of the various kinds of mathematical object. Mathematics is like the natural world in displaying a wide variety of different kinds of object—natural numbers, the reals, points, lines, figures, determinants, groups, and so on. It is therefore natural to seek some kind of uniformity within this diversity. What our theories show is how many of the different kinds of mathematical object can be regarded as forms of abstraction and how their various special theories can be seen to have their basis in the general theory of abstraction. What we achieve is an ontologically significant reduction of certain branches of mathematics to the theory of abstraction, but not an epistemologically significant understanding of the theory of abstraction as a branch of logic.

5. *The Identity of Abstracts*

An abstraction principle yields a necessary and sufficient condition for abstracts of the same sort to be identical. But what of abstracts of different sorts? Given an abstract obtained from an item by one means of abstraction and an abstract obtained from an item by another means of abstraction, under what conditions are the two abstracts the same?

The left-to-right direction of the identity axiom provides a very plausible necessary condition: for the two abstracts to be the same they must be associated, through their respective means of abstraction, with the same equivalence class of items. It seems absurd to suppose that the extension of the concept of being an odd number, for example, could be the number 0; and, in general, it is hard to see what reasonably systematic view could associate the same abstract with two different equivalence classes of items. Perhaps the only view of this sort with any degree of plausibility is the one that identifies higher-order abstracts, i.e. abstracts of abstracts, with lower-order abstracts. It might be supposed, for example, that the non-negative integers mod 2 (obtained from the non-negative integers by means of the equivalence 'leaving the same remainder on division by 2') are the same as the generalized numbers *odd* and *even* (which are obtained directly from the finite concepts by means of a suitable generalization of equinumerosity). But even this view will not yield the cross-sortal identity of *conceptual* abstracts.

Many philosophers have been attracted by the idea that numbers are classes—that a number might be the class of its predecessors, for example, or the class of all concepts that are of that number. If this idea is combined with the view that classes are conceptual abstracts, we would obtain another counter-example to the necessity of our condition. For the very same abstract would, *qua* number, be associated with a range of concepts equinumerous with one another and would, *qua* class, be associated with a range of concepts coextensive with one another; and with possibly the single exception of the number 0, the two ranges of concepts would not be the same. On what basis, then, could an abstract *qua* number and *qua* class be judged the same? The only reasonable view to suggest itself is that *any* abstract, associated through a means of abstraction with certain items, is to be identified with the class of those items. But, as we have seen, such a view leads to the absurd conclusion that any class C of items is identical to the class of the concept (s) whose extension is C.[25]

---

[25] A related objection in Hale (1987–93: 186) is directed against the general thesis that abstracts are to be identified with equivalence classes. He later (p. 187) seems to endorse the view proposed here, that 'abstract objects are differentiated by their association with distinctive [*sic*] equivalence classes'. I therefore find it puzzling that later still, in sect. 8.2, he

The other direction of the identity axiom, though, is much more problematic. Wright (1983: 116–7, 122) and Hale (1987: 206) have discussed the question of sufficiency for cross-sortal identity, but their focus is different from ours. For they want to leave room for the view that numbers are classes; they do not want to reject it out of hand simply on the grounds that numbers and classes are associated with different ranges of concepts. The focus of their discussion is accordingly on the question of how two abstracts might be the same even though they are associated, through their respective means of abstraction, with different concepts.

As I have indicated, this focus appears to be misplaced; for there seems to be no coherent view that will let both numbers be classes and classes be abstracts.[26] Our focus, on the other hand, is on the question of how two abstracts might be distinct even though they are associated, through their respective means of abstraction, with the same concepts. What might contribute to their being different?

The obvious answer is the identity criteria themselves. On this view, then, two abstracts will be distinct if they are associated with different means of abstraction. But even if one does not believe in the

---

should envisage the possibility of numbers being classes even though the equivalence classes associated with a number and with the corresponding class are not the same. It may be that he wants there to allow for the possibility that classes are not a form of conceptual abstract.

[26] I might add that, in my view, neither Wright nor Hale succeed in formulating a *selective* criterion, one that will let in certain identifications and not others. Thus Zermelo numbers are not excluded on Wright's criterion because one can always specify a Zermelo number as the Zermelo number corresponding to a given NBG number (cf. Wright 1983: 123); and Frege numbers are not included on Hale's criterion since one can always specify the Frege number of Cs as the class of Ds for which there exists a many–one correspondence from C onto D and a many–one correspondence from D onto C (cf. Hale 1987: 208). It is important, in this connection, not to confuse the standard terms for certain classes with the classes themselves. Thus Wright (1983: 122) advises us 'to pick on classes among which all questions of identity and distinctness . . . are logically equivalent to questions concerning 1–1 correlation among concepts'. But it is only terms for the classes that can yield the required logical equivalents. Or again, Hale (1983: 208) talks of 'membership of a class' being 'specifiable in terms of 1–1 correspondence'. But the membership of any class whatever might be specified by a formula that uses an expression for 1–1 correspondence and also by a formula that uses no such expression. There is nothing intrinsic to a class, as opposed to a term, that involves 1–1 correspondence.

There are two other aspects of their discussion that one might prefer to be different. One is that they propose a criterion for when all Fs are Gs (e.g. numbers are classes), but one might prefer a criterion for when a particular F is a G. It is perfectly possible, after all, for the Fs and Gs to overlap. The other is that they state the criterion in cognitive terms, such as 'explanation' or 'understanding'. But one might prefer a purely objective formulation. Their more recent discussion of the topic (Hale and Wright 2000*c*: sect. 6) attempts to accommodate these points.

possibility of numbers being classes, this might still appear to be too strong. Suppose one defines *natural number* using equinumerosity as the criterion of identity, but only in application to finite concepts, and that one defines *cardinal number* again using equinumerosity as the criterion of identity, but now in application to all concepts. Does one want to say that the natural number 0 and the cardinal number 0 are not the same on account of their criteria of identity not being the same?

Of course, one might insist in the face of a such an example that a conceptual criterion of identity should have application to all concepts. But suppose that the proposed criterion of identity for natural number is extended to all concepts by treating it as a universal relation on any two infinite concepts. We then obtain a single infinite number ∞, just as in pre-Cantorian mathematics. Do we still want to say that the respective 0s are not the same?

These examples force us to face the possibility that the criteria of identity might be different in a way that is not relevant to the identities of the abstracts in question. And this might lead one to shift to the opposite extreme and take two abstracts to be the same when their associated equivalence classes are the same, regardless of the means of abstraction by which they were obtained. But this view is also subject to difficulties.

For could not two abstracts just *happen* to be associated with the same equivalence class? Perhaps there is a certain weight and a certain density, both unique to a certain thing. Under the proposed criterion, the weight and the density are the same. But they do not appear to be. Indeed, we seem to have a proof that they are not. For surely it is possible for that very weight and density to belong to different things; and if they do, then they will not be the same. (Cf. Hale 1987: 185–6.)

It may be that there does not exist a similar possibility in the case of logical abstracts, such as numbers; for all the possibilities for two such abstracts to be associated with different ranges may, in all essential respects, already be realized. But, be that as it may, one would still like to have a general account of identity, one with application to all forms of abstraction and from which the sufficiency of our original condition, in the case of logical abstracts, might be proved rather than simply assumed.

In the light of these difficulties, it is natural to adopt a modal form of the criterion: two abstracts are to be the same iff they are *necessarily* associated with the same items under their respective means of abstraction. The criterion in effect reduces the intra-world question of cross-sortal identity to the cross-world question of intra-sortal identity. For we can determine whether abstracts of different sorts are the same by 'tracking' each from one world to another and 'seeing' whether the items with which they are associated are always the same.

This question of cross-world identity is clearly important to the development of a modal theory of abstracts. But it is also important to the ontology of abstracts. For suppose we ask: what lengths are there? If one takes lengths to be the lengths of actual things, then there is the danger of missing lengths—a length between two others, perhaps, which we think should exist but which is not the length of anything. On the other hand, if we take there to be lengths that are not the length of anything, we appear to give up on the idea of lengths as a form of abstract.

One natural way out of this dilemma is to take the lengths there are to be the lengths of possible as well as actual objects: given the possibility of a length we suppose there already is that length. But this solution raises the problem (among others) of how the unrealized lengths are related to the realized lengths. On what basis do we say that they are the same or distinct, shorter or longer? Clearly any answer must again rest upon an account of cross-world identity.

So how do we settle the question of cross-world identity? When do we say that an abstract of an item a with respect to a given means of abstraction in one world is the same as an abstract of an item b with respect to the same means of abstraction in another world? One natural line of response is to make a cross-world application of the criterion of identity: the two abstracts are the same if and only if item a in the one world is related by the criterion of identity to item b in the other world.

Unfortunately, a criterion of identity has, in itself, no meaningful application from one world to another; it cannot, Colossus-like, bestride the two worlds. In some cases, though, it is completely clear how the application is to made. We are in no doubt, for example, about the cross-world identity of numbers. The number of Cs in one world will be the same as the number of Ds in another world just in case the Cs and Ds are equinumerous in a world (or domain) in which the respective extensions of C and D are both preserved; and similarly for any other criterion of identity that is logical and absolute in the sense defined above.

Other cases, though, are more problematic. When do we say that something in one world is of the same length as something in another world? Or of the same colour? It seems that any general answer to such a question should be given in terms of the possession of a common property: an item in one world is to be taken to be related by the criterion to an item in another world just in case the one has, in its world, a property of the required sort that is also had by the other in the other world.

Even the idea that we have an alternative explanation of cross-world equinumerosity may be something of an illusion. For we need to make sense of the idea that there is a one-to-one correspondence whose domain is the extension of the concept *C* in the one world and whose range is the extension of the concept *D* in the other world. But if there is no world that contains the two extensions, then why should there be a world which contains the required one-to-one correspondence? Indeed, on any plausible actualist view of the matter, no such world would exist.

Given the need to ground the cross-world identities in properties, what might those properties plausibly be taken to be? A natural suggestion is that they are the relational properties that correspond to the criterion of identity—the properties of being related by the criterion to *a*, for any item *a*. Now it will be true that two items will be related by the criterion just in case they share such a property. But these properties are not, in general, the ones we require. Common possession of such a property is not sufficient, in the cross-world case, for the abstracts to be the same. Two sticks may, in their respective worlds, be of the same length as a given stick or even of the very same things; but that does not mean that they themselves have the same length. Nor is common possession necessary. For one stick in one world may have the same length as another stick in another world and yet the associated equivalence classes in the respective worlds be disjoint.

The relational properties will not even work too well in the case of numbers. If concepts are allowed to have a variable extension, then two distinct numbers may be of the same concept in their respective worlds—they might both, for example, be the number of planets; while if concepts are required to have a rigid extension, there is the problem that they may not exist in every world and so there is the possibility that a number may not attach to a common concept in two given worlds.[27]

Another property that might be associated with an abstract is the property of possessing an abstract of that kind. With a given length, for example, we have the property of being of that length. Common possession of such a property would, of course, be necessary and sufficient for the cross-world identity of the abstracts. But such an account is circular and hence provides no explanation of what the abstracts are. Granted that an explanation is possible, we must therefore have an understanding of the common properties that is not directly explicable either in terms of the criterion of identity or in terms of the abstracts themselves.

In particular cases, we seem to be able to say what these properties are. In the case of numbers, they will be what one might call the *numerical* properties—the property of being a unit concept for example (i.e. of being such that there is an object falling under the given concept to which every object falling under the concept is identical); and in the case of extensions, they will be what one might call the *enumerative* properties—the property, of being a concept of Caesar and Brutus for example (i.e. of being such that an object falls under the given concept just in case it is either Caesar or Brutus). The case of length is more difficult. But one might suppose that significant overlap between paradigm equivalence classes in the two worlds could be used to determine that their items were of the same length. Once a cross-world comparison between these items had been established, they could be used as a yardstick to make the other cross-world comparisons.

As these, and other, examples suggest, there appears to be no uniform method by which the relevant properties can be derived from the criterion of identity. Indeed, it might even be argued that the same intra-world criterion of identity might be associated with distinct cross-world criteria and hence with different kinds of abstract. Consider, for example, the contrast between the conceptions of temperature as given by the Celsius scale and by the thermodynamical account. The concept of same temperature is the same under the two conceptions. But it might be argued that the temperature at

----

[27] I might add that the possibility of the concepts of a given number being different is yet further evidence against the view that a number is the class of concepts of that number, for the membership of a class cannot change.

which water freezes (under normal atmospheric conditions) must remain the same under the Celsius scale but, because of possible changes in the behaviour of water, could be different under the thermodynamical account.

Even when there appears to be only one property type associated with a given criterion of identity, its adoption is by no means forced upon us. In the case of numbers, for example, we could take the property associated with a number to be the property possessed by a concept when it is empty, or singleton, or what have you, *and* the universe of all objects is of the particular size that it is. Equinumerosity would then still serve as the criterion of identity within a world; but in any possible world of a different size from our own, the numbers would all be different.[28]

If this is right, there must be more to our understanding of a form of abstraction than is contained in the adoption of a specific criterion of identity; we must have an understanding of the common properties in virtue of which the identity criterion holds that goes beyond our understanding of the criterion itself. Although it has become standard to identify a form of abstraction with its criterion of identity, it should perhaps be identified with a type of property, common possession of which provides the basis for the criterion to hold.[29]

In the simplest case, and perhaps in general, the properties will be mutually exclusive and exhaustive over the domain of items on which they are to be defined. Thus with each item x, we may associate a unique property P_x of the type in question. If R is the corresponding criterion of identity, then the two should conform to the condition:

$$(*) \quad x\mathbf{R}y \text{ iff } P_x = P_y$$

for any items x and y. If we think of the relation $\mathbf{R}$ and the properties P_x in extensional terms, then the choice of P_x, for any given x, is uniquely determined by $\mathbf{R}$; and it is perhaps this fact that has made it so easy to overlook the independent role of the properties P_x. However, if we think of the relation and the properties in intensional terms and require that (*) hold of necessity for any items x and y, then

---

[28] Tennant (1994), in a different way, has stressed the independent importance of numerical properties in understanding what numbers are; and Peacocke (1991) has stressed the general importance of properties in providing a modal account of our understanding of abstracts.

[29] I now feel less confident of this conclusion and am inclined to think that it may be possible to meet the objections to the relational account.

the choice of P_x is by no means uniquely determined by **R**. For each way of drawing the transworld-lines between the equivalence classes in different worlds will correspond to different choices of P_x.

In developing the theory below, I have only been interested in the extensional properties of abstracts and therefore have had no reason to depart from the conventional approach. However, it seems to me that, within a fuller account, the means of abstraction should perhaps be specified in terms of property types rather than equivalence relations. The theory of classes, as developed above on the basis of the theory of abstracts, might be seen as a step in this direction—though, of course, the Extensionality axiom must be given up.[30]

Let me mention, in conclusion, that one might envisage a criterion of identity that was intermediate in strength between the present modal criterion and the earlier criterion in terms of the means of abstraction. Under the modal criterion, the number 0 and the empty class will be the same; for in each possible world they will be associated with the same concept (s), namely the empty one (s). However, one might wish to maintain that the two are distinct and that therefore something more is required for the identity of two abstracts than their necessarily being associated with the same range of items.

I myself would be inclined to question the intuition that lies behind this view. For to the extent that we wish to conceive of a class as an abstract on concepts, it is not clear that the null class is to be distinguished from the number 0. But if the intuition is upheld, it is hard to see how it might coherently be developed. For the relevant difference between the number and the class seems to be tied to the type of abstract in question. What is common to the empty concepts can either be regarded as a form of number or as a form of extension. But if the two can be distinguished in this way, then what is to stop us distinguishing between the natural numbers that are tied to a single pre-Cantorian infinity and the ones that are tied to the usual range of infinite numbers? To draw the line, we seem to require an, as yet, undeveloped theory of basic classificatory types or forms.

---

[30] Under an intensional conception of abstracts as envisaged here, there is a new danger of inflation; for the possibilities of distinguishing abstracts associated with the same actual range of items might exceed the number of items in the universe. This difficulty must therefore also be resolved by a satisfactory intensional account.

II

The Context Principle

IT was once a common part of mathematical practice to introduce terms for new entities by means of contextual definition. Thus negative numbers might have been introduced by means of the equation, $-m + m = 0$; and the point at infinity might have been introduced by the condition that $r < \infty$ for any real number r. Such definitions seem to hold out the hope that one might be justified in taking there to be certain entities simply upon the basis of a stipulation. Thus we may conclude on the basis of the stipulation of $-m + m = 0$ that there is a solution to the equation $x + m = 0$; and we may conclude on the basis of the stipulation of $r < \infty$ that there is a 'number' greater than every real. If this idea could be extended to the more basic objects of mathematics, such as the natural numbers themselves or to the reals, then it might seem as if a large part of mathematics could be seen to be the product of stipulation and that, consequently, many of the most pressing problems concerning the meaning and justification of mathematical statements could be bypassed.

The 'context principle' can be regarded as an attempt to vindicate such contextual definitions. It was first advocated by Frege in the *Grundlagen* and, although he himself gave up on the attempt to introduce natural numbers by contextual definition, this project has subsequently been revived by such neo-logicists as Wright (1983). I wish in this part to see how the principle should be formulated and whether it can be used to provide a contextual definition of number. My interest, as with the topic of reconceptualization, is more in the ideas themselves than in any question of Fregean exegesis.[1]

We first attempt to provide a preliminary account of what the context principle is (sect. 1). The discussion is then organized around

---

[1] The reader might like to consult C. Parsons (1964), Wright (1983), Hodes (1984), and Dummett (1981*a*; 1981*b*; 1991*a*; and 1991*b*) for general discussion of the context principle.

various features that it might be thought desirable for contextual definitions to possess. These are that they should yield truth-conditions for all of a suitable range of sentences (sect. 2), that they should conform to a solution to the Caesar problem (sect. 3), that they should yield a determinate assignment of referents to the terms that are to be defined (sect. 4), that they should not be viciously circular in the sense of quantifying over the very entities that are to be assigned as referents to the terms (sect. 5), and that they should yield a satisfactory account of number (sect. 6).

Although the upshot of my discussion is largely negative, I am hopeful, as I have indicated in the Preface, that there may be a more satisfactory way of achieving the benefits that the context principle appears, so tantalizingly, to place within our grasp.

1. What is the Context Principle?

Implicit or explicit definitions of a standard sort are made from a standpoint in which the existence of the objects or items that are to be assigned to the defined terms is presupposed. The purpose of the definition is not to introduce new objects into the domain but to make an appropriate assignment of the objects already in the domain to the terms that are to be defined. Thus prior to the definition being made, we should be sure that the required objects from which the assignment is to be made already exist if, after the definition has been given, we are to be sure that these objects have been appropriately assigned to the terms.

What I call creative definitions, on the other hand, are made from a standpoint in which the existence of the objects that are to be assigned to the terms is not presupposed. The purpose of the definition may indeed be to assign objects to the terms. But these objects are not selected from a previously given domain. Rather the objects are introduced into the discourse simultaneously with their assignment to the terms.

Thus, if all goes well, the usual order of justification will be reversed. With implicit definition of an orthodox sort, we know that the definition is referentially effective by knowing, prior to the definition being made, that objects of the required sort exist. But with a creative definition, we may know that objects of the required sort exist by knowing that the definition has been referentially effective. The definition itself may provide us with the ground, or at least part of the ground, for supposing that objects of the required sort exist.

But how can creative definition work? How can a mere definition or stipulation, made from a position of ontological neutrality, be of any assistance in providing us with reference to or knowledge of something new? There are different answers to this question that might be given. One, that we have already considered and found lacking, is that the definition may proceed through the reconceptualization of a given content. The answer on which I now wish to focus and which, I believe, is far more worthy of consideration is that such definitions may proceed via the context principle (CP).

The general idea behind the principle, as I understand it, is that linguistic practice may be partly constitutive of reference. The fact that certain terms are used in a certain way may guarantee, in conjunction with the appropriate non-linguistic facts, that those terms refer and that they refer to what they do. The practice need not be taken to create the objects or items, for it may be supposed that they exist independently of our practice of referring to them; but it will help create or constitute our reference to those items. The practice will be constitutive of our reference to the objects, if not of the objects themselves.

Now there is a way in which such a view might be completely unexceptional. For we might take it to be part of the use of the terms that they refer to what they do; or we may include within the use of the terms the fact that they have an identifying sense of the usual sort. Thus in neither of these cases need there be anything exceptional in the way the terms are taken to refer.

If the view is therefore to be of any interest, it must be that there is some other, less obvious, way in which linguistic practice can conspire with the world to deliver reference. It must be supposed that the relevant use of the terms does not explicitly involve the fact that the terms refer to what they do or that they have an identifying sense of the usual sort. Let us call the linguistic behaviour of terms that can be characterized without directly imputing either reference or an identifying sense to them their 'apparently referential behaviour'. Then the view is that the apparently referential behaviour of terms may help secure their reference: terms that behave *as if* they refer *will* refer, given that the appropriate non-linguistic facts are in their favour.

In general, the apparently referential behaviour of the terms will not by itself secure their reference. For reference in this case, just as in the standard case in which there is an identifying sense, is a co-operative endeavour, requiring a contribution both from language and from the world. What will be true, though, is that the apparently linguistic behaviour will provide all that is required on the linguistic side; as long as the world is co-operative, and all the appropriate non-linguistic facts obtain, reference will be secured. We might say that the apparently linguistic behaviour of the terms endows them with the *potential* to refer, a potential that will be realized when the appropriate non-linguistic facts obtain. Thus referential potential is the counterpart, within the framework of CP, to an identifying sense within the standard Fregean model of reference.

But what might plausibly be taken to constitute the apparently referential behaviour of a term? What short of reference itself or of an identifying sense might plausibly be taken to secure the potential to refer? Three main aspects of our linguistic practice have been appealed to in this regard. The first is that the terms should behave in the same syntactic fashion as referential terms, i.e. of terms that refer or purport to refer. The second is that they should be subject to the same logical principles as referential terms. And the third is that the sentences that contain them should be subject to appropriate conditions of truth and falsehood, those that we would expect to obtain if the terms did indeed refer.

It is natural to take each of these condition on use to be implied by its successor. For how can the logic of the terms be of the right sort unless their syntax is? And how can the truth (and falsehood) conditions be of the right sort unless their logic is? Thus the third condition embodies the full requirement: the referential potential of the terms is to be given through the truth-conditions of the sentences in which they occur.

Given the Fregean conception of sense, it is natural to take the truth-conditions of a sentence simply to be its sense. The view would then take the referential potential of the terms to be given by the sense of sentences that contain them. But the context principle need not be tied in this way to a Fregean conception of sense. We could imagine, for example, that the truth-conditions were stated within a theory of truth of the sort advocated by Davidson and that no assumption was made concerning either the existence or assignment of sense.

Nor need it be maintained that the relevant use of the terms must be explained exclusively in terms of the truth-conditions of the sentences that contain them; appeal might also be allowed to other ways in which the terms or sentences are used. Consider the case of a logical constant, such as '&'. A sentence of the form 'P & Q' will be true just in case P and Q are true. But in order to avoid the problems of circularity to which such a formulation gives rise, one might want to say instead that the relevant use of '&' consists in our willingness to infer 'P & Q' from P and Q and to infer each of P and Q from 'P & Q'.

However, in what follows I shall always suppose that the relevant use of the terms to which CP is to be applied can always be stated in the form of truth-conditions. The cases in which one might be tempted to think otherwise will be of no interest to us; and I suspect that they involve considerations of a very different sort.

Among the truth-conditional applications of CP, there is a further distinction to be drawn between those cases where the terms can be understood *by means of* a statement of the relevant truth-conditions and those cases where it cannot. Thus it has sometimes been supposed that one might understand the use of terms for abstract objects through a specification of the truth-conditions for the sentences that contain them. No special experience or recognitional capacity is required; one need merely understand the statement by which the truth-conditions are given. On the other hand, it is sometimes supposed that our understanding of proper names for concrete objects requires an ability to recognize the bearer of the name, in addition to the acceptance of a criterion of identity for the kind of object in question. But such an ability cannot simply be extracted from a statement of truth-conditions; it would normally also require some kind of contact with the real world.

In the first case, the statement of the truth-conditions may be used to provide a definition of the term or terms in question. The statement will provide one with a purely linguistic route to an understanding of the terms. But in the second case, no such route is available. The truth-conditions can perhaps be stated; but their formulation will require the use of terms whose understanding is already in question. One might attempt to minimize the differences between the two cases, either by playing up our contact with the bearers in the abstract case or by playing it down in the concrete case. However, in what follows, I shall assume that our understanding of the abstract terms of interest to us—such as those for number or direction—does not rest upon any form of non-linguistic contact with their bearers; and I shall be concerned to see to what extent CP might then be taken to account for our understanding of these terms.

2. Completeness

Let us try to be more precise as to what is required for the successful application of CP. One important requirement—that the specification of the truth-conditions should be complete—is taken up in the present section. A significant aspect of completeness, the Caesar problem, is then discussed in the following section; and a significant possible consequence of completeness, referential determinacy, is discussed in the section after that.

It will be helpful here (as with definitions of a more orthodox sort) to distinguish between two rather different kinds of situation in which the context principle might be applied. On the one hand, the terms to be defined might belong to a previously understood language; and our aim, in applying CP, might be to provide a definition of those terms that is in conformity with our existing understanding of them. All cases of philosophical interest are of this sort. On the other hand, the terms to be defined might not already be understood; and our aim, in applying CP, might be to provide ourselves with an understanding of those terms. Since a situation of the first kind can always be understood by reference to a hypothetical situation of the second kind in which we succeed in recovering our actual understanding of the terms, nothing will be lost by confining our attention to the second case.

Let us suppose, therefore, that we are given a language L whose terms and sentences are already understood and that we add new terms to L to form the (as yet only partially interpreted) language L^+. It is our aim in applying CP to L^+ to secure reference, or at least potential reference, for the new terms of L^+; and we do this by adding a set of 'bridge principles' B, which relate the truth (and falsehood) of sentences in L^+ to the truth (and falsehood) of sentences in L. We might call this set of bridge principles a *contextual definition* or *definition by CP* for the new terms.

Thus any purported application of CP will be given by three components: the given language L; the extended language L^+; and the bridge principles B. But not all choices of these components will constitute an acceptable application of CP: given the base language L, not all extensions L^+ will be considered admissible; and given an extension L^+, not all choices of bridge principles will be considered admissible. Which selections of L, L^+, and B might legitimately be made is a large question upon which it is difficult to attain a secure view.[2]

There are, however, two key conditions that are, at least, necessary for a successful application of CP. The first—which we discuss here and label 'completeness'—is that truth-conditions should be provided for all sentences of the extended language L^+ that contain any of the new terms. Or perhaps we should be more cautious. For where a new term t does not have a referential occurrence (as with the sentence '"t" is a term') or when the resulting 'sentence' is not meaningful, then we do not expect an application of CP to provide truth-conditions. Thus what we require is that truth (and falsehood) conditions should be provided for the sentence $\phi(t)$ as long as: (a) there is a meaningful notion of what it is for an object (or for an object of an appropriate sort) to satisfy the condition $\phi(x)$; and (b) the term t may be meaningfully substituted for x in $\phi(x)$ without alteration to its logical form.[3] For we certainly expect the sentence to have truth-conditions in such a case; and if they are not provided by CP, there would appear to be no other way in which they might be given.

It should be noted that we do not expect CP to tell us whether such sentences *are* true or false but merely to provide the *conditions* for their truth or falsehood. We also do not insist that those conditions be such as to guarantee that *every* sentence actually be true or false. For the aim of an application of CP is to provide the new terms with referential *potential*; and when the world is unkind, this potential may not be realized. If one is of the view that sentences containing non-referring terms lack a truth-value, for example, then CP should be taken to deliver a truth-value gap in such cases. But even here, the principle should provide us with the conditions under which the sentence lacks a truth-value; it cannot simply remain silent.

It is not necessary that the application of CP should *directly* specify the truth-conditions of every meaningful sentence that contains a (referential) occurrence of one of the new terms. For some of these

---

[2] This question is also discussed by Hale and Wright [2000*b*] within the general context of implicit definition.

[3] Fine (1989; 1990) contain an extensive discussion of how this latter condition might not be met.

conditions may be given indirectly, as consequences of the directly specified truth-conditions and of general rules of use. The most notable case of this sort is provided by the logical constants. Thus, given the conditions of truth and falsehood for the sentence ϕ, the conditions of truth and falsehood for the sentence not-ϕ may be determined according to the following two rules:

N(i) not-ϕ is true iff ϕ is false; and
N(ii) not-ϕ is false iff ϕ is true.

Similarly, the truth- and falsehood-conditions for a disjunction are given by:

D(i) (ϕ or ψ) is true iff ϕ is true or ψ is true (and perhaps also that neither is truth-valueless); and
D(ii) (ϕ or ψ) is false iff ϕ and ψ are both false.

The provision of rules for the quantifiers is not so straightforward. Consider, first, the case of first-order quantification. Given that the first-order quantifiers in L range over a given domain of objects, we shall take the first-order quantifiers in L^+ to range over the objects from that domain plus the referents of the new terms. We could leave the range of the quantifiers alone. But the applications of interest to us are ones in which we shall wish to quantify over the objects introduced by CP; and, in any case, if referents for the new terms are indeed secured, then it should be possible to quantify over a domain of objects that includes them and so any reasonable application of CP should tell us how such quantification is to be understood. Whether the referents of the new terms should be already allowed to appear in the given domain (i.e. as values of the first-order variables of L) is a question we shall consider later, though the natural assumption is that they should not.

In order to simplify the formulation of the rules, let us suppose that each object from the given domain has a name in L, that the newly introduced terms are exclusively terms for objects, and that we have the means of determining whether any given term refers (a question we take up at the end of the section). We might then determine truth-conditions for universally quantified sentences in accordance with the following pair of rules:

Q(i) $\forall x \phi(x)$ is true iff $\phi(t)$ is true for each (closed) referring term t; and

Q(ii) $\forall x\phi(x)$ is false iff $\phi(t)$ is false for some referring term t (and perhaps also that $\phi(t)$ is never truth-valueless for a referring term t).

However, for certain applications, the above rules will not lead to a well-grounded account of the truth-conditions and, for this reason, it may be preferable to select a range of *privileged* terms to represent the objects of the domain. The first rule then takes the form:

Q(i)$'$ $\forall x\phi(x)$ is true iff $\phi(t)$ is true for each privileged referring term t;

and similarly for the second rule. It should be noted that there does not appear to be any reasonable way in which the truth-conditions for quantified sentences might be understood in terms of satisfaction, since the induction must eventually terminate in the specification of truth-conditions for *sentences*.

Second- and higher-order quantification raise further problems. We might give a substitutional rule for quantification over concepts of objects, in the manner of Q(i) above:

CQ(i) $\forall c\phi(C)$ is true iff $\phi(F)$ is true for each (closed) one-place predicate-term of L^{+}.

But it is well known that, when the predicate-term F itself contains second-order quantifiers, the resulting truth-definition may fail to be grounded. If the proponent of CP is prepared to be platonistic about quantification over concepts and not assume that each concept must be specifiable by means of a predicate, then he should presumably also be willing to be platonistic about concepts of terms. Thus each concept may, in effect, be identified with a corresponding set of terms; and quantification over concepts can be construed as substitutional quantification with respect to the possibly infinitary predicate-terms of the form $\lambda x(x = t_1 v x = t_2 v \ldots)$, where $t_1, t_2, \ldots$ are referring terms from L^{+}. Given that these predicate-terms are used in CQ(i) in place of the less explicitly given terms F, the problems over grounding can be finessed. (This approach is pursued in sect. III.6 and shown to lead to well-defined truth-conditions.)

If, on the other hand, the proponent of CP is not prepared to be platonistic about higher-order quantification, then he may well adopt an account of how it is to be construed that can be adapted to the present case. If, for example, he 'ramifies' the conceptual quantifiers in *L*, there should be no difficulty in extending the treatment to the conceptual quantifiers in L^+. Thus in neither case does higher-order quantification appear to pose an additional problem for the proponent of CP.[4]

If the truth-conditions of logically complex sentences can all be indirectly determined in this way, then all that remains, in order to satisfy the completeness requirement, is to determine the truth-conditions of the atomic sentences. If the provision of the truth-conditions for the sentences of L^+ does not involve any new predicates, then the new atomic sentences of L^+ will be of the form $Pt_1t_2 \ldots t_n$, where P is an old predicate from L and at least one of the terms $t_1t_2 \ldots t_n$ is new.[5] Now it is conceivable that not all of the old predicates will have meaningful application to the new terms. If one of the old predicates is 'is Roman', for example, and one of the new terms is '3', then one may wish to deny that the sentence '3 is Roman' is meaningful. This was not Frege's mature view; but if it is adopted, then not all applications of the old predicates to the new terms need be eligible for the assignment of truth-conditions.

The case of identity, however, is special. For whatever one's view on the other predicates, it is plausible that the identity predicate should have meaningful application to any terms whatever. Given that the two terms t and s putatively refer, then surely the sentence 's = t' will be meaningful and hence subject to conditions under which it can properly be said to be true or false. Moreover, should both the terms s and t refer, the identity-sentence s = t itself will be either true or false. For the terms either refer to the same object, in which case the sentence is true, or to distinct objects, in which case the sentence is false. Thus the truth-conditions assigned to the identity-sentences in these cases should result in a definite assignment of truth-value.

---

[4] Thus I am not sure that I would agree with Hazen (1985: 252–3) that 'the impredicative notion of *concept* implicit in his [Wright's] acceptance of second-order logic is if anything more contentious than that involved in object impredicativity'. The case of second-order quantification may, however, accentuate a problem he already has. For the formulation of the truth-conditions in a metalanguage requires that we presuppose an infinite ontology of abstract expression-types; and matters are only made worse when one quantifies over arbitrary sets of such expressions.

[5] The antecedent condition is not automatically satisfied. If, for example, infinitary 'enumerative' predicates are used to interpret second-order quantification, then these must be allowed to figure as the predicates in atomic sentences. I assume, though, that newly introduced predicates of this sort will not give rise to any special problems.

The only way in which this conclusion might plausibly be resisted is to suppose that the reference of the terms s or t is indeterminate, either in the sense that it is not determinate to what they refer or in the sense that what they refer to is not determinate. There are two relatively innocuous ways in which this might happen. It is possible, on the one hand, that the application of CP has only been partially specified and that under different ways of rendering it complete the terms might enjoy different referents. And it is possible, on the other hand, that one or both terms are complex and that the indeterminacy has an independent source from within the term itself. Thus the vagueness of the predicate 'bald' might be taken to render the reference of 'the number of bald men' indeterminate and thereby deprive the sentence 'the number of bald men in the room = 3' of a definite truth-value.

But let us suppose that there is no external source of indeterminacy (as with the predicate 'bald') and that the specification of the given application of CP is complete. Can a term introduced through CP still have indeterminate reference? This would require that there be a truth-value gap in an identity-sentence involving the new term, since otherwise there would be no reason to suppose that its reference was indeterminate; and it would also require that the gap be incapable of being closed, since otherwise there would be no reason to think that the application of CP was complete. It is conceivable that the proponent of CP might be willing to admit cases of this sort, for he may wish to avoid the paradoxes (that arise for extensions and the like) by positing the existence of irresolvable gaps in truth-value. But if such cases are admitted, they are best seen as exhibiting the second form of indeterminacy. For within the framework of CP, we seem incapable of forming a coherent conception of the objects between which the new terms might be indeterminate. For these objects would themselves have to be given to us through the application of CP; and, by hypothesis, there are no further objects of this sort. With this one possible exception, then, 'autonomous' gaps in the truth-value of identity-sentences should not be allowed. We may countenance a truth-value gap for the sentence 'the number of bald men in the room is 3' on account of an indeterminacy in what it is to be bald, but not on account of an indeterminacy in what it is to be the number 3.

There are two kinds of identity-sentence for which truth-conditions should be provided:

(i) those of the form s = t, where both s and t are new terms;
(ii) those of the form s = t, where one of s or t is a new term and the other is not.

We may call sentences of the first sort *unmixed* and those of the second sort *mixed*. Thus, in the arithmetical case, the unmixed cases are those of the form 'the number of Fs = the number of Gs', where F and G are predicates from the extended language L^+, and the unmixed cases are those of the form 'the number of Fs = t' or 't = the number of Fs', where t is a term that is *not* of the form 'the number of Gs' and F is, as before, a predicate of L^+.[6]

The truth-conditions for the unmixed cases will often reveal what is most distinctive about the objects, or terms, in question. This will be true of terms for Fregean numbers, for example, and indeed of all terms for Frege-style abstracts: identity will 'translate' into the appropriate form of equivalence. But this is not generally true of contextual definitions. What is most distinctive about the symbol '∞' for the point at infinity is that 'r < ∞' is true for any real term r. Or again, what is most distinctive about a term '−n' for a negative integer is that the sentence 'n + (− n) = 0' should be true (here identity is involved, but not an identity of the form '−m = −n').

The problem presented by the mixed identities 's = t' is to say when an object designated in one kind of way is identical to an object designated in some other kind of way. If it is apparent from the manner of designation what *sort* of object is designated, then this is the problem of determining when an object of one sort is identical to an object of some other sort. The famous example of Frege's is 'the number of F's = Caesar' (*Grundlagen*, sect. 56). Thus, granted that 'the number of Fs' designates a number and that 'Caesar' designates a person, the rule for assigning truth-conditions to mixed identities in the arithmetical case must somehow have as a consequence that no person is a number.

In addition to assigning truth-conditions to mixed identity-sentences, an application of CP should also assign truth-conditions to all other mixed predications $Pt_1t_2 \ldots t_n$, where P is an atomic predicate from the given language L and at least one of the terms $t_1, t_2, \ldots, t_n$ is new. We need to say in the case of numbers, for example, under what conditions the number 3 is Roman (granted, of course, that it may

---

[6] Unless the given language L itself contains operation-forming terms, in which case a term might be novel through containing an embedded occurrence of the number-operator.

meaningfully be said to be Roman). We might call this the 'Roman' problem, by analogy with Frege's Caesar problem.

It might be thought that the Roman problem is simply a special case of the Caesar problem. For the sentence '$\forall x(\text{Roman}(x) \leftrightarrow x = \text{Caesar} \vee x = \text{Brutus} \vee \dots)$' will be true in L; and so given that each of '3 = Caesar', '3 = Brutus', ... is false, it will follow that 'Roman (3)' is false. But this is to overlook the fact that, within L, the quantifier '$\forall x$' may only range over the objects countenanced by L and not also over the objects introduced in L^+. Thus we may not be justified in substituting '3' for 'x' in the universal claim above; and if we are not, then some other way of ascertaining that 3 is not a Roman must be found. What is perhaps true is that the Roman problem, though not reducible to the Caesar problem, may be solvable by essentially similar means; for as long as the Roman-type predicates are sortally circumscribed, the solution in either case may be taken to turn on the question of when an object of one sort is identical to an object of another sort.

Before turning to the question of how the Caesar and Roman problems might be solved, let us take up a question left open by our treatment of quantification. For this required that we be able to distinguish between the terms that refer and those that do not. How is this to be done?

There is a straightforward answer to this question. For quite apart from CP, we recognize that the truth (or falsehood) of certain sentences requires that certain of their terms refer. A term can then be taken to refer if there is some sentence whose truth requires, in this way, that the term refers and whose truth-conditions, as determined by the application of CP, are satisfied. Indeed, such a view is forced upon the proponent of CP. For he must acknowledge that the term refers in such a case; and within the framework provided by CP, there would appear to be no other basis upon which he could take a term to refer.

Which truths can be taken to guarantee reference is a general issue within the philosophy of language; and the proponent of CP is not obliged to adopt one position rather than another. On the standard Fregean view, no sentence can be true or false if it contains a term that does not refer; and so any truth or falsehood containing a given term can be used to secure reference. Under other views, any atomic sentence containing a non-referring term will be false; and so only certain of the truths can be used to secure reference.

If it is granted that truth-conditions for identity sentences (or, at least, self-identity sentences) should be laid down and also that no identity sentence can be true unless its terms refer, then the criterion for reference can be given, in a simple uniform manner, in terms of identity. For the term t can be taken to refer just in case the truth-conditions for t = t are satisfied; the reference-conditions for a term are simply the truth-conditions for the corresponding sentence of self-identity.

Thus, in keeping with the general spirit of CP, the usual order of explanation is reversed. Instead of saying that the sentence 't = t' is true in virtue of the fact that the term t refers to a given object x and that object is identical to itself, we say that the term t refers in virtue of the fact that the sentence 't = t' is true—where this latter fact is itself explained without appeal to the reference of 't'.

3. The Caesar Problem

I wish in this section to propose a schematic solution to the Caesar problem (and the corresponding Roman problem). We shall see that although the proposal applies in a relatively straightforward manner to definitions by abstraction, its application to other forms of contextual definition sanctioned by CP is somewhat problematic. The main other philosophers to have attempted a positive account are Wright (1983: sect. vii), and, more recently, Hale and Wright (2000c), but their approach is rather different from mine, despite some obvious points of contact.

A solution to the Caesar (and Roman) problem should be formally adequate in the sense of helping to settle the truth-value of each meaningful 'mixed' atomic sentence. It should, that is to say, provide truth-conditions for such sentences which, in conjunction with the 'facts', will determine whether or not they are true or false. The solution should also be in conformity with our intuitions, i.e. not deliver any incorrect results; and it should possess the usual explanatory virtues (non-circularity, systematicity, etc).

For ease of exposition, let us confine out attention to the Caesar problem (concerning identities) since most of what we say about this problem will straightforwardly generalize to the Roman problem (which concerns all other atomic predications). Now it might be thought that the Caesar problem could simply be solved by fiat. For suppose we lay down the following two requirements: first, contextual definitions should be linearly ordered (with a first, a second, etc.); and second, any contextual definition for a new term must indicate how it relates to all antecedent terms. Then any acceptable system of contextual definitions will already embody a solution to the Caesar problem; since, as the language expands, truth-conditions will be assigned to every meaningful sentence that has so far been formed.[7]

It might be objected that this proposal merely shifts the problem to another place. For the question now is: what constitutes an acceptable contextual definition? Suppose I introduce Hume's Law in a language that already contains the term 'Caesar'. Then according to the proposal, I must also stipulate, either directly or indirectly, whether or not Caesar is the number 3. But then what is to stop me from taking him to be the number 3?

The answer is 'nothing'. But that is not because the resulting definition would constitute an acceptable account of number. It would be acceptable—though not as an account of our pre-existing notion of number, but of a strange variant of that notion, call it 'cumber', in which the office normally performed by the number 3 is now performed by Caesar. Thus it might be maintained that we can resolve the questions of mixed identities in any manner that we choose. The resulting definition may or may not be in conformity with some pre-existing notion but it will not, in either case, give rise to any special Caesarian concerns.

A different objection is that the solution to the Caesarian problem should be implicit in a definitional 'core'—Hume's Law or what have you—without the addition of any special stipulations concerning the truth-conditions for mixed identities. However, under the present proposal, the same core might belong to different definitions—of number, say, or cumber; and so it is not clear why the core should be taken unequivocally to fix the notion to be defined.

The most serious objection to the proposal, I believe, is that the requirement that the definitions be linearly ordered is unduly restrictive (or that it is, once it is coupled with the further requirement that the definition should take account of all antecedent terms). What is to stop one from defining number by Hume's Law, or the like, and *independently* defining direction by the corresponding principle of abstraction for directions? It is not that one comes first or the other

---

[7] This seems to be the idea behind Frege's application of CP in *Grundgesetze* (1893–1903: i. sect. 10).

second; and so there is no need for either definition to address the question of whether any number is the same as a direction.

Even if it is insisted that the linearity requirement be met, a similar problem will arise for definitions from *different* languages. Suppose that number is defined by Hume's Law in one language and that direction is defined by a corresponding principle of abstraction in another language. Then neither definition will make reference to the other; and so, again, there will be no evident basis upon which we can settle the question of whether the objects defined by the one definition are the same as the objects defined by the other. Thus a cross-linguistic form of the Caesar problem will arise, even if the strictures on the proper form of definition prevent it from arising within a language.

We therefore cannot in general expect a solution to the Caesar problem to follow explicitly from the contextual definitions themselves. But how then is the problem to be solved? How should the definitions be supplemented or how should they be read, if the solution is not already explicit in the definitions?

In addressing this question, it will be helpful to divide the Caesar problem into two parts: we need to say when a mixed identity is *true*; and we need to say when it is *false*. Now suppose that we have a complete account of when a mixed identity is true. We may then declare a mixed identity to be false if it cannot be shown to true (assuming Bivalence, of course). In other words, if the identity does not follow from the account, the relevant contextual definitions, and the underlying facts, then it may be deemed to be false. We might call this the 'completeness rule', since its valid application presupposes the completeness of the positive account.[8] Given the completeness rule, our task simply reduces to the task of saying when a mixed identity is true.

It might be thought that, once armed with the completeness rule, a simple solution to the Caesar problem is once again available. For why not simply abstain from providing any positive account of which mixed identities hold? In other words, why not accept an account in which the maximum of mixed identities are taken to be false?[9] It will

---

[8] Such a rule is familiar from other contexts: it corresponds to the 'extremal' clause in inductive definitions and to the 'closed world' assumption of non-monotonic logic.

[9] Not all mixed identities need be false, since the truth of some may follow from the contextual definitions themselves. Thus it should presumably follow from the contextual definition for negative integers that $-0 = 0$.

then follow that Caesar is not the number 3, simply because it does not follow from the contextual definition (along with the 'facts') that Caesar *is* the number 3.

From a formal point of view, this is indeed a simple and elegant solution. The question is whether we can be content with its results. For might there not be identities between terms from different contextual definitions that we feel obliged to accept?

I believe that there are. Perhaps the least controvertible cases concern what one might call 'duplicate' definitions. Suppose I define 'number' by Hume's Law: the number of Fs = the number of Gs iff F and G are equinumerous. (Nothing here turns on the choice of Hume's Law; any other contextual definition of any other notion would do in its place.) Suppose now that I independently define 'qumber' by Hume's Law: the qumber of Fs = the qumber of Gs iff F and G are equinumerous. It does not then follow from the two forms of Hume's Law that any number is the same as any qumber— that the number of non-self-identicals, for example, is identical to the qumber of non-self-identicals; and so we should conclude, under the minimalist application of the completion rule, that numbers are not qumbers. Yet surely they are the same and surely, in particular, the number of Fs is identical to the qumber of Fs for any concept F.

My opponent might simply dispute our intuitions on this point and insist that numbers and qumbers, as so defined, will *not* be the same. But then in what does the difference between numbers and qumbers consist? And given that they are not the same, how come the definition in terms of 'number' serves to pick out numbers while the definition in terms of 'qumber' serves to pick out qumbers? How can the use of one definiendum as opposed to another make a difference to which objects are defined? It might be supposed that each definition indeterminately picks out both numbers and qumbers; under any admissible precisification of either predicate, the referents will be either numbers or qumbers (or rumbers etc). But we have now lost our motivation for distinguishing between numbers and qumbers and, even if we take there to be both numbers and qumbers, it is not altogether clear why we should not maintain that numbers are qumbers, since this will be true under any 'co-ordinate' determination of what we might take the referents to be. Just as the indeterminacy of 'number' is no bar to the truth of 'numbers are numbers', so the indeterminacy of 'number' and 'qumber' need be no bar to the truth of 'numbers are qumbers'. (Even if the referents of 'number' and 'qumber' are not co-ordinated, we still require an explanation as to why statements of the form 'the number of Fs = the qumber of Fs' are neither true nor false.)

Our opponent might also object that, in the application of the completion rule, we have operated with too narrow a notion of what it is for one thing to follow from another. Certainly, it will not *logically* follow from Hume's Law for 'number' and 'qumber' that numbers are qumbers, but it will follow *analytically*, since it is an analytic truth that numbers are qumbers.

I think this point may be conceded. But part of what is here in question is what we should take the analytic truths to be. Given a definition of number and direction in terms of the relevant abstraction principles, it may well be true, and hence an analytic truth, that numbers are not directions. But in attempting to solve the Caesar problem, we cannot take for granted that it is an analytic truth; for part of our aim should be to account for how the analytic import of contextual definitions is capable of extending beyond their logical import.

There therefore appears to be no alternative but to go beyond the minimalist approach and provide a more positive account of which mixed identities should be taken to hold. Let us therefore take duplicate definitions, ones ascribing the same logical content to their respective definienda, to give rise to identical referents. We should go further still. For suppose I provide a contextual definition of numbers mod 2 in terms of numbers in the usual way (the number mod 2 of n = the number mod 2 of m iff m and n have the same remainder upon division by 2); and let us suppose that we provide a duplicate definition of qumbers mod 2 in terms of qumbers. Then if numbers are identified with qumbers, numbers mod 2 be should identified with qumbers mod 2. Duplicates at one level should be taken to give rise to duplicates at a higher level; and the referents of duplicates at the higher level should be taken to be the same.

But is this very modest departure from minimalism enough? Or are there non-duplicate forms of definition that should also be seen to give rise to identical referents? If our only concern is with contextual definition via Fregean abstraction, then the present proposal may well be adequate. The proposal amounts to saying that Fregean abstracts are the same only when their underlying equivalence relations are the same. Thus even if we define *cardinal number* using Hume's Law and *natural number* using the principle that the number of Fs = the natural number of Gs iff F and G are both infinite or F is equinumerous with G, it will not be possible to identify the finite natural numbers with cardinal numbers, since the underlying equivalence relation is not the same. Although the view is subject to some obvious reservations, which we discussed in sect. I.5, it is not clearly incorrect; and the reservations, for what they are worth, seem to have more to do with getting straight on what cross-sortal identities should be taken to hold rather than with seeing how they might be accommodated within a theory of contextual definition.

The main reason why this simple proposal has not been accepted is that it does not allow room for the view that cardinal numbers, and the like, might turn out to be classes or some other kind of object. Suppose that 'number' is defined by Hume's Law and 'extension' by a suitably circumscribed version of Law V.[10] One might then wish to claim that each number was identical with the extension of the corresponding second-order concept, even though this did not follow from the two laws under the identification of duplicates. Similarly, if one had a prior notion of set (not necessarily given by contextual definition), then one might wish to identify the number of a set with a von Neumann-style cardinal. Perhaps these are not the views to have, but a solution to the Caesar problem should at least allow them to be on the cards.

I find this response excessively tolerant. From the general standpoint of CP, there is something highly suspicious about a term that is *capable* of being introduced by means of a contextual definition *also* having an explicit definition. For how could the provision of truth-conditions for sentences containing number-terms, let us say, lead one to suppose that those number-terms should themselves have an explicit definition in terms of some other term-forming operator (unless this is already evident from the truth-conditions themselves)? It is hard to see how such an understanding of the contextual definition could be permissible unless we were in general free to adopt any interpretation of the number-terms that was compatible with the stated truth-conditions. But then that would be to abandon the whole idea of CP as a distinctive form of definition and simply to treat it as a standard form of implicit definition. If this is right, then a contextual definition of *number*, as given by Hume's Law for

---

[10] A consistent version of Law V that enables one to define cardinals à la Frege and then derive Hume's Law is given in sect. I.4.

example, is simply not compatible with any form of explicit definition.

Even if 'external' interpretations of a contextual definition are allowed, we still face the problem of excluding incompatible interpretations. If, for example, the number of Fs may be identified with the class of all concepts equinumerous with F, then why should it not also be identified with the class of all concepts that are equinumerous with F or some subconcept of F? In each case, it will follow, as a matter of logic, that Hume's Law is satisfied; and there appears to be no other ground upon which we might reasonably discriminate between the two identifications. But the number of Fs cannot be both the one class and the other; and so it cannot be either.[11]

Whatever its merits might otherwise be, the view that numbers are classes is incoherent within the context of CP; and the attempt to accommodate it within a solution to the Caesar problem can only lead us astray. So can the present proposal, perhaps subject to some fine-tuning over the identity of abstracts, be allowed to stand? I think not; and this is because contextual definition via Fregean abstraction is not the only form that contextual definition may take. It is worth bearing in mind, given the huge emphasis on Fregean principles of abstraction, that there is nothing in the general idea of CP that limits its application to such principles. Indeed, the historically most important examples of contextual definition concern the various extensions of the number system; and these are not most naturally taken to be given by abstraction. Thus what is important about negative numbers is not that $-m = -n$ iff $m = n$, but that $-m + m = 0$; and what is important about rational numbers is not that $m/n = p/q$ iff $m.q = p.n$ but that $m/n.n = m$. In the case of the imaginary number i or the 'point at infinity' ∞, there is not even an abstraction principle in the offing: i is essentially governed by the principle that $i^2 = -1$ and ∞ by the principle that, for any real number r, $r < \infty$.

Once we attempt to deal with this wider class of contextual definitions, I think it will be seen that the present solution to the Caesar problem is seriously deficient.[12] Consider the extension of the system

---

[11] A further reason, internal to the theory of abstracts, for not identifying the number of Fs with the class of concepts equinumerous with F was given in sect. I.2.

[12] Let me mention two other cases in which preoccupation with definition by abstraction may have led to over-generalization. Hale and Wright (2000*b*: sect. 3) seem to think that implicit definition cannot proceed by outright categorical stipulation, but the case of

of natural numbers to positive and negative rationals. We can imagine doing this in two different ways. We might first introduce the
negative integers (essentially through the equation $-m + m = 0$)
and then introduce the rationals (essentially through the equation
$m/n.n = m$). On the other hand, we might first introduce the rational numbers and then the negative numbers. Let us suppose that
we independently stipulate both pairs of definition and, to keep
notation straight, let us use '$-$' and '$/$' for the first pair of definitions
and their boldface counterparts '$-$' and '$/$' for the second pair. Now
surely we will wish to say that $-1/2$ and $-(1/2)$ are the same
number. Indeed, we appear to have the same compelling reasons as
in the case of duplicate definitions for considering them to be the
same. However, on the present proposal, they will not be the same:
for the respective definitions '$-$' and '$-$' and of '$/$' and '$/$' are not
duplicates, since they are given over different domains, and so there
will be no basis for taking their referents to be the same.

A similar problem arises over the definition of the points of
infinity, $-\infty$ and $+\infty$. One might introduce $-\infty$ first and then
$+\infty$, or $+\infty$ first and then $-\infty$, or both of them simultaneously
(subject to the condition that $-\infty < r < +\infty$ for any real r). Intuitively, the objects defined in each case are the same and yet, according
to the proposal, they are different.

On what basis can we say that the objects defined in these various
cases are the same? It seems to me that two main things are involved.
The first is to identify certain structural relations, it being the point of
a contextual definition to specify the structural relations that hold
between the objects to be defined and the given objects. Thus in the
case of the integers and the rationals, the structural relations are given
by the arithmetical operations of addition and multiplication while,
in the case of $-\infty$ and $+\infty$, the sole structural relation is $<$. The
second is to say what structural relationships are implicit in any given
definition. Thus we may introduce $+\infty$ as being greater than all reals
or as being greater than all reals *and* $-\infty$ (if $-\infty$ is introduced first).
But we want the two positive points of infinity to be the same and so
it must somehow be implicit in the first definition that $-\infty > +\infty$,
even though no reference is explicitly made to $-\infty$.

∞ (which is stipulated outright to be greater than every real) shows otherwise. They also
seem to think (2000c: sect. 7) that there is no special problem over cross-categorical
classification; but the case of $-1/2$, which should be classified both as a negative number
and as a rational, creates difficulties for this view.

A great deal more needs to be said on the matter, but let me merely point out that the framework provided by CP may be well be unsuited to the task of specifying the relevant structural relationships. The main difficulty arises from the fact that CP characteristically permits one to introduce many objects simultaneously. I might introduce the two points of infinity, ∞_1 and ∞_2, for example, subject to the condition that $r < \infty_1 < \infty_2$ for any real r. Suppose that I now introduce a single point of infinity ∞ subject to the condition that $r < \infty$. Then on what basis am I to identify ∞ with ∞_1 or ∞_2, if with either? The problem does not arise if the objects to be defined are introduced one at a time, since then the first of ∞_1 or ∞_2 to be introduced can be identified with ∞. But if we subject CP to the requirement that it only introduce one term (and hence one object) at a time then we greatly curb its scope and, in particular, all the standard definitions by abstraction will have to be abandoned. It therefore appears as if one will need to make a radical break with the framework in which contextual definitions are usually set if the Caesar problem is to be solved.

Let me conclude this section by noting an intimate connection between the Caesar problem and a doctrine concerning our access to contextually defined objects. According to this doctrine, which we may call 'Limited Access', the means by which a contextually defined object is introduced into the discourse provide essentially the only means by which it may be identified. Thus if numbers are introduced by Hume's Law, then any particular number must essentially be identified as the number of a given concept; and, similarly, if directions are introduced by a corresponding principle of abstraction on lines, then any direction must essentially be identified as the direction of a given line.

It is not clear exactly what this doctrine amounts to. One difficulty is to say what is meant by 'essentially'. Consider again the case of numbers; and suppose they are introduced via something like Hume's Law (the precise details will not matter). Suppose now that t is a term for a number (i.e. that it refers to a number). Then it would be going too far to insist that t must be of the form 'the number of Fs' for some predicate F. For it might be of the form 'the x such that x is the number of Fs'. It would also be going too far to insist that t be analytically co-referential with a term t' of the form 'the number of Fs' (i.e. that the sentence 't = t'' be analytic). For the term t might be of the form 'the x such that x is identical to 0 if snow is white and is identical to Caesar otherwise'. Since the sentence 'snow is white' is true but not analytic, the term will refer to the number 0 without being analytically co-referential with any term of the form 'the number of Fs'. However, in all the above cases the mechanism for determining the reference of the term in accordance with its sense involves identifying the referent as a number; in executing the procedure that corresponds to the sense of the term, we must invoke a 'subroutine' in which the referent is determined as the number associated with some concept. This there-fore suggests that a term cannot be a term for a number unless the semantic mechanism for determining its reference requires identify-ing the referent *as* a number.

The other difficulty is to say what is meant by 'means'. Suppose, starting with the natural numbers, we first introduce the negative integers and then introduce the rationals. Thus $-1/2$ is introduced as the ratio of -1 to 2. But it might also be identified as the negative of $1/2$. It is therefore clear that, in identifying a given contextually defined object, we should allow ourselves to use whatever structural relations might be directly or indirectly involved in formulating the definition.

With these two provisos in place, it is then plausible that some-thing like the doctrine should hold. And what this means, in effect, is that the solution to the Caesar problem should be of a suitably minimalist sort; objects should be identified only in so far as the structural relations by which they are given can be identified.

Strictly speaking, the doctrine of Limited Access is not about numbers or directions or the like but about the objects, whatever they might be, that can be introduced by means of contextual defini-tion. But it is plausible that a related doctrine should hold for abstract objects as ordinarily conceived, and not necessarily as the product of contextual definition. For our access to abstract objects such as numbers or sets or directions appears to be limited by the structural relations with which they are naturally associated. A minimalist solution to the Caesar problem should therefore provide an explan-ation for this fact: since it will then follow, as long as these objects are capable of being contextually defined, that no other means of identi-fying them will be available to us.

4. *Referential Determinacy*

We turn to the question of how a contextual definition might be capable of achieving referential determinacy. I shall suggest three main ways in which this might be done, the first two relying on completeness and the third on the idea of canonical reference. Although I think that each of these approaches may be capable of achieving determinacy, it is at the expense of reducing contextual definition to a standard form of definition; and so the apparent epistemic advantages of adopting CP are lost.

Suppose that we attempt to introduce the negative integers by means of the following contextual definition:

$$-n + n = 0.$$

Then it is natural to suppose, if one accepts the legitimacy of such a definition, that it serves to fix the reference of the negative terms '$-n$'—that, once given the definition, there can be no real question as to what the negative numbers are or how they are to be assigned to the negative terms. And it might be supposed, in general, that any legitimate definition by CP will also serve to fix the reference of its terms.

But how is such determinacy of reference achieved? If we simply regarded the specification of the relevant truth-conditions as an implicit definition of a standard sort, then they would be compatible with the assignment of almost any objects as referents of the terms to be defined. In the definition of negative numbers above, for example, the negative numbers $-1, -2, \ldots$ could be any objects whatever as long as they were all distinct from one another and from the positive integers and as long as the operation of addition was so understood as to render $-n + n$ always equal to 0. Thus the negative numbers could be points or lines, or even cabbages and kings. How then is a contextual definition able to secure referential determinacy?

The requirement of completeness to some extent alleviates the problem. Suppose we insist that truth-conditions should be provided, either directly or indirectly, for every meaningful statement that contains the terms to be defined. Then negative numbers cannot be cabbages or kings, since the general principles governing their contextual definition (e.g. the completeness rule of sect. 3) will somehow rule this out. However, even with the completeness requirement being met, some indeterminacy will remain. For one thing, we may arbitrarily permute the assignment of the negative numbers $-1, -2, \ldots$ to the negative terms '-1', '-2',$\ldots$ and still satisfy the stipulated truth-conditions, as long as the arithmetical operations are suitably reinterpreted. In the second place, we may arbitrarily switch negative numbers with directions, say, or classes, again subject to reinterpretation of the various basic operations. Thus within the whole domain of contextually defined objects, there will essentially be no constraints on which objects may be assigned to which terms.

How then is this residual issue of determinacy to be resolved? There are, I believe, several answers that might, with some plausibility, be given. According to one view, once we have a complete assignment of truth-conditions, then no intelligible doubts can be raised concerning the determinacy of the definition. There are two versions of this view. The first is disquotationalist in character.[13] It states that any meaningful question we may raise in our language concerning whether a given term 't' refers to s (e.g. '3' to Caesar) simply reduces to the questions of whether the term 't' is meaningful and of whether t = s. But this means that once we have resolved all questions of identity, the only question that remains is whether the term is meaningful; and it is plausible that, for a disquotationalist, nothing more than the full provision of truth-conditions would be necessary to render a term for an abstract object meaningful. Thus once the completeness requirement is met, any doubts concerning determinacy will disappear.

The second version of the view rests upon adopting a form of holism. It may be granted that the possibility of reinterpretation is normally indicative of referential indeterminacy. But it will be maintained that this is only true when the reinterpretation is not of a general systematic kind. If every truth (or all truth-conditions) are preserved by the reinterpretation, then no sense can be given to there being a genuine difference in reference. The objects assigned to the terms only enjoy a relative identity *vis à vis* one another and, if a difference in their relative identity is not manifest in the truths within which they figure, then no genuine difference can be attributed to what they are.

This version of the view might be understood by way of an analogy with location. Normally, a difference in the assignment of position to things will correspond to a genuine difference in the locational facts. But it might be maintained that even though local variation in position is possible, there is no genuine possibility of systematic variation. Thus even though there is a possibility of my chair being

---

[13] Dummett (1991*a*: 155–6, 192; 1991*b*: 39) attributes such a view to the Frege of the *Grundlagen*; and it has recently been advocated and developed by Field (2001).

2 feet to the left of where it is (keeping the relative position of everything else fixed), it is not possible that everything might be 2 feet to the left of where it is. Or rather, what appears to be a different possibility is merely a different representation of the very same locational facts. And similarly, what might appear to be a different referential assignment is merely a different representation of the same underlying referential facts, given that the truth-conditions remain the same.

There is a third, rather different, way of attempting to secure determinacy. It may be allowed that a contextual definition, no matter how complete it might be, is capable of being satisfied by genuinely different assignments of referents to its terms. However, it will be maintained that a contextual definition does not merely demand that the referents conform to the stated conditions, but that they conform in a special way. There must be no more to the referents, so to speak, than what is required for them to satisfy the given conditions. Thus the integer -3 is the *mere* solution of the equation $x + 3 = 0$; and it is on this account that it gets picked out by the equation (considered as a contextual definition) in preference to a cabbage or a king.

It is hard to say, in more basic terms, what this special relationship between a term and its referent is meant to be. One might compare it to the relationship between a term and itself when the term is being used autonomously. Although some other term might be used to refer to that term, there is a specially intimate way in which that term refers to itself. Of course, the object introduced by a contextual definition is not the same as the defining term; it is a kind of objectified version, or shadow, of the term. But it might be thought to stand in a specially intimate relationship to the term which, though not as transparent as identity, is equally definitive of what the object is.

The present view posits a special definitional mechanism whereby referents are to be assigned to terms; and it is the canonical character of the referents, rather than the completeness of the definition, that accounts for the possibility of determinacy. For this reason, there is no need for a contextual definition to provide a complete specification of the truth-conditions, even implicitly. The role of a solution to the Caesar problem is not to complete an otherwise incomplete definition but to make clear what follows from the special way in which the defining conditions are to be satisfied.

Although both views have some plausibility, it has to admitted that neither is altogether satisfactory. The difficulty, in each case, is to maintain what is distinctive about contextual definition and to avoid collapse into a standard form of implicit or explicit definition. Under the first view, it is hard to see why contextual definition should not be regarded as a special case of implicit definition. It only differs from the normal cases of implicit definition by being more systematic; and what guarantees determinacy is not some special definitional mechanism but some general limitations, of either a disquotationalist or holistic sort, on what is referentially possible. Under the second view, on the other hand, it is hard to see why contextual definition should not be regarded as a form of explicit definition. For why should we not import the notion of canonical conformity into the object-language and then explicitly define *number-of*, let us say, as that operation which canonically conforms to Hume's Law?

But then what comes of the apparent epistemic advantage of CP? If a contextual definition is really an implicit definition, then objects of the required sort must be shown to exist if it is to be legitimate; and if it is a covert form of explicit definition, then objects of the required sort must be shown to exist if it is to be of any use. So the idea that we might have achieved referential determinacy without the usual epistemic cost is still without justification.

5. Predicativity

If we are to set up a contextual definition for certain terms, then we need to state truth-conditions for the sentences that contain them or, to state the matter more carefully, we need to indicate what contribution the presence of the terms will make to the truth-conditions of all those sentences that do indeed possess truth-conditions. But not any formulation of the conditions will do. We cannot take the truth-conditions for a sentence about number, when numbers are what is in question, to be given by the sentence itself. For the specification of the truth-conditions should be non-circular; it should make no use of the terms to be defined or in any other way presuppose an understanding of them.

This is not to say that these terms cannot be used at an *intermediate* stage in the formulation of truth-conditions. In providing the truth-conditions for a general statement, for example, we may appeal to instances that involve the very terms in question, as long as the use of these terms is eventually eliminated. If we think of the specification of truth-conditions as providing us with an inductive procedure, in which one statement of the truth-conditions may give way to another, then there will be two ways in which the specification may fail to provide us with a non-circular outcome: it may terminate in a statement, or in statements, of truth-conditions that already presuppose an understanding of the terms to be defined; or it may not terminate in a statement of truth-conditions at all, either because it goes round in a circle or because it results in an infinite regress.

If the above criterion for non-circularity is to be made precise, we need to be able to say when a statement provides an appropriate terminus in an account of truth-conditions. When is a statement not question-begging, i.e. when does it not presuppose an understanding of the terms whose definition is in question? In attempting to answer this question, it will be useful to distinguish between three kinds of statement:

(1) *Statements of Grade 1*—that do not contain the terms in question nor quantify over (or otherwise appeal to) the referents of the terms;

(2) *Statements of Grade 2*—that do not contain the terms in question but do quantify over (or otherwise appeal to) the referents of the terms;

(3) *Statements of Grade 3*—that contain the terms in question.

In the number case, for example, the statement that all men are mortal will be of grade 1, the statement that every object is self-identical will be of grade 2, and the statement that the number of planets is 9 will be of grade 3. We might put the difference in terms of the distinction between *definite* and *indefinite* reference. Grade 1 statements involve neither definite nor indefinite reference to the questionable objects, grade 2 statements involve only indefinite reference, while grade 3 statements involve definite reference (and possibly indefinite reference as well).

It will be generally agreed that grade 1 statements are not question-begging and that grade 3 statements are. This leaves grade 2 statements; and it is here that controversy lies. For although these statements do not invoke the problematic notions (number-of, extension-of, etc.), they do appear to invoke a problematic ontology (numbers, extensions, etc.) Does the non-circularity of a contextual definition require merely that we avoid the problematic notions or also that we avoid the problematic objects? We take *predicativism* to be the view that grade 2 statements *are* question-begging, and *impredicativism* to be the view that they are not. Thus a predicativist will not allow himself to quantify over the referents of the terms to be defined in providing truth-conditions for statements involving those terms, while the impredicativist will allow such quantification.

The two views lead to very different conceptions of contextual definition. According to the predicativist, when we define certain terms by CP we enlarge the domain of quantification; we actually introduce into the domain of quantification objects that were not previously there. For the impredicativist on the other hand, the domain of quantification will remain the same under a contextual definition; what happens when we define certain terms by CP is that we single out certain objects that were previously undifferentiated. Let us distinguish between our indefinite referential ken, which consists of the objects we can quantify over at a given stage in our understanding of a language, and our *definite* referential ken, which consists of the objects to which we can make definite reference. For the predicativist, contextual definition enlarges our indefinite referential ken while, for the impredicativist, it merely enlarges our definite referential ken (the objects are put into sharper referential focus, as it were). Contextual definition is genuinely ontologically innovative on the one view, and merely referentially innovative on the other.

Given these distinctions, there are three critical questions we should ask. First, in any case in which we wish to give a contextual definition, can we provide a non-circular account of the truth-conditions, even of an impredicative sort? A negative answer to this question would be highly significant since it would show that the impossibility of a contextual definition could be established without even engaging the issue of predicativism. But let us suppose that the answer is positive and that an impredicative definition can be found. We may then ask: is predicativism a legitimate requirement on any contextual definition? If it is not, then the previous impredicative definition can stand. If it is, then we should consider the further question of whether a predicative account of the truth-conditions can also be given. I shall consider the first two questions in the present section and turn to the last question in the next section.

We consider first the question of when it is possible to provide an impredicative account of the truth-conditions. Let us take the proposed definition of number by means of Hume's Law as a typical example of an impredicative account (we shall later consider how our discussion of this example might be generalized). We shall assume that the quantifiers in the formulation of Hume's Law range over all objects (and concepts) whatever, since it is hard to see how the inclusive quantification in terms of which the truth-conditions might be stated under an impredicative approach could be anything other than unrestricted quantification. Let us also suppose that we have a solution to the Caesar problem. This then should presumably provide us with a general understanding of the number-predicate 'x is a number' (though not, of course, of the number-operator 'the number of—').

Now let A be an arbitrary arithmetical statement, one that may contain the number-of operator and may involve unrestricted quantification over all objects. Then can we, on the basis of Hume's Law, assign to A an impredicative account A′ of its truth-conditions, where A′ is a statement that, like A, may involve unrestricted quantification over all objects and that may also contain occurrences of the number-predicate, but will not contain any occurrences of the number-operator? In other words, can the number-of operator be eliminated in favour of the number-predicate?

I suspect that the answer has usually been taken to be 'no'. For it has been supposed that even though Hume's Law allows us to eliminate the general statement 'the number of Fs = the number of Gs' in favour of a non-arithmetical statement of one-to-one correspondence, it does not allow us to eliminate general statements of the form 'x = the number of Fs'; and this is so even if we are provided with the additional information that x is a number.[14]

The correct answer, however, is 'yes'. For consider an existential statement of the form:

(*) for some x, A(x).

This is equivalent to a disjunction of:

(*1) for some non-number x, A(x); and
(*2) for some number x, A(x).

The second of these, in its turn, is equivalent to:

for some F, A (the number of Fs).

---

[14] Wright (1997): sect. 1; 1998*a*: n. 15), for example, claims that Hume's Law does not provide the resources to eliminate the occurrence of the operator N in $\exists y(y = Nx{:}x \neq x)$.

If we make substitutions in accordance with these equivalences, then every statement of the form 'x = the number of Fs' can be taken, within a given context, to be false since it will have been assumed within that context that x is not a number. We are therefore left only with identities of the form 'the number of Fs = the number of Gs', which can be eliminated on the basis of Hume's Law.[15] Thus we have shown how effectively to associate with each arithmetical statement A an impredicative statement A′ (not containing the number-operator) for which it provably follows from Hume's Law (and the definitional principle that x is a number iff it is the number of Fs for some F) that A is equivalent to A′.

Although the number-operator is eliminable in favour of the number-predicate, it is not *definable* in terms of the number-predicate. In other words, there is no formula A(x, F), not involving the number-operator, for which:

x is the number of Fs iff A(x, F)

is a consequence of Hume's Principle.

The above reasoning in favour of a negative answer establishes, at best, that the number-operator is not definable.[16] But, as we have seen, this is still compatible with its being eliminable.

The use of arithmetical vocabulary 'on the right' cannot be avoided altogether, since there is no purely logical equivalent for the statement that there is at least one object that is not a number. However, the total elimination of arithmetical vocabulary *is* possible in certain special cases. Say that a statement is *purely* arithmetical if it is arithmetical and if all the objects that it refers to or quantifies over are restricted to numbers (and similarly, if all concepts, relations, etc. are restricted in their application to numbers). In this case, we may dispense with the first disjunct (E′) above; and so any purely arithmetical statement $\phi(N)$ will be provable equivalent (given Hume's Law) to a purely logical statement. Thus when it comes to

---

[15] Thus the formula $\exists y \, (y = Nx: x \neq x)$ from the previous footnote will be equivalent to $\exists F \, (Nx:Fx = Nx:x \neq x) \lor \exists y(- Ny \, \& \, y = Nx:x \neq x)$ which, in its turn, is equivalent to $\top$, which is just what one would expect. I have assumed that A contains no non-logical constants other than the number-operator. If it contains other non-logical constants, then they can perhaps be handled on the basis of a general solution to the Caesar problem (or the corresponding Roman problem).

[16] I say 'at best', since it merely excludes one obvious way of defining the number-operator. What really establishes indefinability is the 'switching' argument of Lemma III. 3.8.

the statements of arithmetic itself, it is possible to give a completely 'clean' account of their truth-conditions.

Or again, let us suppose that ψ is a statement saying how many objects there are that are not numbers.[17] Say that an operation O taking all concepts into objects is *Humean with respect to* ψ if it conforms to Hume's Law and if the complement of its range (the 'non-numbers') is of the cardinality specified by ψ. Suppose now that $\phi(N)$ is a statement that contains occurrences of the number-operator N. Then it follows from Hume's Law and the cardinality statement ψ that $\phi(N)$ is equivalent to the statement:

(**) $\phi(O)$ holds for some operator O that is Humean with respect to ψ.[18]

Thus once we know the cardinality of the non-numbers, we are in a position to specify the truth of every arithmetical statement in purely logical terms.

Moreover, these results hold for a wide range of abstraction principles. The first (in terms of (*)) holds for any abstraction principle whose criterion of identity is not flatly circular; while the second (in terms of (**)) holds for any abstraction principle that is absolute and invariant.[19] We may therefore conclude that, for a wide range of abstraction operators, it is possible to provide an impredicative account of the truth-conditions for statements containing the operator *as long as* the corresponding abstraction predicate can be used in formulating the truth-conditions or *as long as* the cardinality of the class of non-abstracts is taken to be known.

Let us now turn to the second question of whether the impredicativist approach is legitimate? Can we, in providing a contextual definition of numbers or directions or the like, quantify over the

---

[17] The statement $\psi(D)$ should specify the cardinality of D in the sense that the formula $\forall D, E[\psi(D)\ \&\ \psi(E) \supset (D$ is equinumerous with E$)]$ should be a logical theorem (or a logical truth).

[18] The proof is as follows. Supose $\phi(N)$ holds, then $\phi(O)$ holds for some Humean operator O with respect to ψ, since N is Humean with respect to ψ. Now it may be shown that any two Humean operators with respect to ψ are 'isomorphic', so that if $\phi(O)$ holds for some Humean operator O with respect to ψ, then it holds for any other Humean operator with respect to ψ and, in particular, for N. Note that the equivalent is not found, in this case, by substituting the left-hand side of Hume's Law for the right-hand side.

[19] In the sense of sect. III.3. I might add that the third-order quantification over operators can be avoided by using the trick described in sect. IV. 1.

very objects that are in question?[20] Imagine someone who has not yet been exposed to a contextual definition of number. Can he be taken to have a prior understanding of quantification over a domain of objects that includes the numbers themselves and in terms of which the truth-conditions for statements concerning number might then be framed?

I am inclined to answer this question on doctrinaire grounds: since the point of a definition by CP is to introduce a certain ontology of objects, we should not appeal to that ontology in explaining how the objects are to be introduced. But I believe that there might also be a less doctrinaire basis upon which a negative answer can be given. For the impredicativist owes us an account of the truth-conditions for those statements whose quantifiers range over the inclusive domain. It is not merely that this is a reasonable demand in itself, it is required by his general strategy for providing truth-conditions for the statements containing problematic terms. For this proceeds inductively, with the truth-conditions for logically complex statements being explained in terms of the truth-conditions for their simpler components. But then, for this strategy to work, the impredicativist must specify truth-conditions for statements of the form $\forall x \phi(x)$, where $\phi(x)$ contains one or more of the problematic terms; and so he must also be in a position to specify truth-conditions for such statements, even when they do not contain a problematic term.

How are the truth-conditions for the statements of the form $\forall x \phi(x)$ to be specified? Presumably, in terms of their instances. Again, this is not merely a reasonable demand in itself but would appear to be required by the inductive strategy, since it is otherwise unclear how the induction is to proceed. Among the instances of $\forall x \phi(x)$, given that the quantifiers are inclusive, will be ones that concern the objects that are the referents of the terms to be defined. Now there would appear to be two ways in which the relevant instances might be conceived—either objectually or substitutionally. In the first case, the truth of $\forall x \phi(x)$, with respect to the relevant instances, will require that $\phi(x)$ be true of every one those objects; and in the second case, its truth will require that $\phi(t)$ be true for each of the terms t that is to be defined (or perhaps for a suitable selection of such terms).

---

[20] I should emphasize that my question entirely relates to the piecemeal, as opposed to the global, approach to contextual definition. Impredicativity is not merely acceptable, but *essential*, on the holistic approach since what is required, in effect, is a vast implicit definition.

Consider each case in turn. The first requires that we give an account of what it is for an object to satisfy an open statement $\phi(x)$ and hence will require, in particular, an account of what it is for an object to satisfy an open statement 'x = t', where t is a 'generic' term for one of the new objects. Thus, in the case of number, we must say, with respect to any concept *C*, what it is for an object to satisfy the open statement 'x = the number of Cs' (if the quantifiers over the objects are objectually interpreted, then so presumably are the corresponding quantifiers over the concepts). But this will provide us with what is, in effect, an explicit definition of the number-of operator; for if $\Psi(C,x)$ is the condition for an object x to satisfy the open statement 'x = the number of Cs', we may define the number of Cs as the x for which $\Psi(C,x)$. We might attempt to evade this conclusion by appealing to the notion of satisfaction only for the numbers. But this requires that we be able to single out the objects that are numbers; and so, given that we know what it is for a number to satisfy the open statement 'x = the number of Cs', we will know what it is for an arbitrary object to satisfy the open statement.

The second case requires that we provide an account of what it is for a closed instance $\phi(t)$ to be true, where t is one of the new terms, and so, once the induction unravels, it will require that we provide an account of the truth-conditions for statements of the form t = t'. But the account will then fail to provide determinate truth-conditions, since the truth-conditions for the identity statements will be given in terms of quantified statements and the truth-conditions for quantified statements in terms of the identity statements. Thus neither approach will work; we end up with either too much or too little.

One might question this line of reasoning on the grounds that it may not be necessary to resort to a standard inductive strategy in specifying the truth-conditions. Indeed, the impredicative accounts of truth-conditions stated above do not pursue such a strategy but proceed in terms of a wholesale transformation of the given statement. So what, in particular, is to prevent the impredicativist from adopting such an account?

The reason, I think, turns on the use that might properly be made of a solution to the Caesar problem. For the solution will not provide us an account of what it is for an arbitrary object to be a number; it will not, in other words, provide us with an explicit definition of number. Rather it will provide us with truth-conditions for mixed identities of the formed t = t', where t and t' are both closed terms, but one is of the form 'the number of Fs' while the other is not. But knowing the truth-conditions of such statements will not allow us to eliminate arbitrary occurrences of the number-predicate; and so the use of the predicate in the formulation of the truth-conditions will not be justified after all. Indeed, it appears that the only way a solution to the Caesar problem might be of any use in formulating the truth-conditions is as part of a standard inductive strategy. For how else are we to produce the mixed identities upon which a solution to the Caesar problem can then be brought to bear?

The second account (in terms of the cardinality specification Ψ) might appear to avoid any reliance upon a solution to the Caesar problem. But this is not really so. For we will need such a solution to ascertain the cardinality of the objects that are not numbers. Moreover, the account will break down once further non-logical constants are added to the language; and again, we will need to bring in the solution to the Caesar problem in order to ascertain what further conditions should be placed upon a Humean operator if the account is still to work. It therefore appears that, in this case too, there is no reasonable alternative to the inductive strategy. Thus even if one is open, in principle, to allowing an impredicativist account of the truth-conditions, there would appear to be difficulties in integrating such an account with a solution to the Caesar problem.

As I have said, I am inclined to endorse predicativism on doctrinaire grounds, and so impredicative truth-conditions should not be allowed even if they *could* be reconciled with a solution to the Caesar problem. But it has to be admitted that the doctrinaire endorsement of predicativism leaves the predicativist in an embarrassing predicament. For surely, it might be thought, it is possible to form the conception of *unrestricted* quantification prior to any application of CP; and if this is possible, then why should we not employ such quantification in stating the truth-conditions of any given contextual definition?[21]

In responding to this predicament, the predicativist might simply deny that we can form an intelligible conception of unrestricted quantification. But this is a hard line to take; and it would be better if the predicativist were not forced to take it.

---

[21] This is a tack taken by Wright (1998*a*: 391–4). For further discussion of this issue see Wright (1998*b*: 353–4), Dummett (1963; 1991*a*: ch. 24; 1993: 429–45) and the first three papers in Brandl and Sullivan (1998).

But is there an alternative? Let us distinguish two different ways in which an understanding of unrestricted quantification might be prior to a given contextual definition. It might be prior in the sense that we can understand the quantification independently of the contextual definition; or it might be prior in the stronger sense that it can legitimately be used in formulating the truth-conditions for that definition. The predicativist might then concede that we can have a prior understanding of unrestricted quantification in the first sense without conceding that we can have it in the second sense. He might allow, in other words, that there was a prior understanding of unrestricted quantification that was of no help in formulating the truth-conditions.

But the question might now be pressed: if we can form an understanding of unrestricted quantification prior to a given contextual definition, then why should it not be used in stating the truth-conditions? What is the relevant difference between the partial understanding that we acquire of unrestricted quantification after a contextual definition is made and the understanding we have before it is made? Intuitively, there is a difference. For our understanding prior to the definition is purely schematic, we have no real conception of what the objects are or how they might differ from other objects; and it is only once we have made the definition that we are able to form a full-blooded conception of what the objects are. However, the predicativist needs to make clear what this distinction is and why it matters. I believe this can be done but let us, for now, merely note that this problem is one that the doctrinaire predicativist must solve if his position is to be rendered coherent.

6. The Possible Predicative Content of Hume's Law

I turn to the question of finding predicative truth-conditions and I wish to consider, in particular, whether Hume's Law might be capable of providing us with a predicative understanding of number, even though the principle itself is impredicative in form. I argue that previous attempts to provide Hume's Law with predicative content are subject to severe difficulties and that a fundamentally different approach to the problem is required if these difficulties are to be solved.

It is clear, in principle, that a contextual definition that is impredicative in form might be capable of providing us with a predicative understanding of the term or terms to be defined. Consider the contextual definition of a negative integer $-n$ as an integer for which:

(*) $-n + m = p$ iff $m = p + n$.

Now the defining condition is meant to have application to the case in which n, m, and p are negative integers and is therefore impredicative in form. However, it is straightforward to extract a predicative meaning from this definition. For we can restrict (*) to the case in which n, m, and p are non-negative integers and then give an explicit specification of how (*) is to be extended to the other cases. So, for example, when m is also a negative integer, say $-k$, we can stipulate that $-n + m = -(n + k)$.

Another, somewhat different, example is with the contextual definition of ordered pair:

(**) $<x, y> = <u, v>$ iff $x = u$ and $y = v$.

Since the variables x, y, u, and v are meant to range over ordered pairs, the definition is also impredicative in form. However, we might provide it with a predicative reading in the following way. We suppose that there are names for all *individuals*, i.e. for all objects in the domain that are not ordered pairs. We construct terms for ordered pairs (pairs of such pairs etc.) from these names in the obvious way and then subject the language with quantification over ordered pairs to a substitutional semantics, using closed instances of (**) to provide the truth-conditions for identity statements between ordered pairs. Since (**) results in a reduction of rank (defined as the maximum degree of embedding of the ordered pair operator), the truth-conditions will eventually bottom out in statements that merely concern the identity of individuals. In this case, in contrast to the previous one, there is no obvious way of factoring out the principle into an explicit definition, on the one hand, and a component that is obviously predicative, on the other. The predicative content is somehow read into the principle by means of a substitutional interpretation of the quantifiers.

Our question is whether anything similar is possible in the case of Hume's Law (and of other such principles). In considering this question, it is important to bear in mind that any acceptable predicative reading of Hume's Law will provide it with additional logical content, i.e. will render statements true that are not themselves logical consequences of the Law. And here I have in mind not merely

Caesar-type statements (such as 'Caesar is not a number') but also statements of pure arithmetic, in which the only non-logical constant is the number operator and in which the quantifiers are restricted to numbers. For the predicative content of an arithmetical statement can only turn upon how many individuals there are. But the truth-value of the statement 'the number of natural numbers = the number of numbers' does not get settled by Hume's Law, even when the cardinality of the class of individuals is taken to be fixed. Since the statement *should* receive a truth-value on a predicative reading (given the underlying predicative facts), such a reading must somehow take us beyond the logical content of the Law itself.[22]

Two questions now arise. Which truth-values *should* be assigned to the arithmetical statements under a predicative reading (given the predicative facts)? And *how* should they be assigned? What, in other words, is the mechanism by which the predicative reading assigns the truth-values that it does? There would appear to be only one reasonable answer to the first question. The truth-values should be assigned on the basis of the minimalist interpretation of sect. I.2 in which the numbers are successively generated from the individuals and from previously generated numbers. For given that the predicative reading should assign a truth-value to purely arithmetical statements (on the basis of the predicative facts), there would appear to be no other reasonable basis upon which this might be done. Any other choice would be completely arbitrary.[23]

How the truth-values are to be assigned is a more difficult matter. Is it possible that each arithmetical statement might be translated into one that exactly describes its predicative content (in analogy to the impredicative translation of the previous section)? In case the domain of individuals is taken to be infinite, such a translation can indeed be given. Say that an operation O is *quasi-Humean* if it is an operation that takes concepts of individuals into individuals in such a

---

[22] Boolos (1987: 16) and Dummett (1991*a*: 227) have observed that the truth-value of the statement 'the number of objects is identical to the number of natural numbers' does not get settled by Hume's Law. But the present form of incompleteness is more radical in that it concerns a purely arithmetical statement and still holds even when given all the non-numerical facts. I might also add that it is not subject to the objection canvassed by Wright (1998*b*: 401) and Hale (1994: sect. 6) that we should not expect to assign a number to all objects since *object* is not a genuine sort.

[23] We might note that the earlier substitutional reading of the contextual definition for ordered pairs also yields a minimal model and also yields additional logical content (at least in a second-order setting), since every ordered pair is required to be well-founded.

way that (1) Hume's Law is satisfied (within the domain of individuals) and (2) the individuals in the range of the operation are equinumerous with those that are not. Thus, with a quasi-Humean operation, 'half' of the individuals go proxy for numbers. Given any statement $\phi(N)$ whose quantifiers range over both individuals and numbers and which may contain occurrences of the number-operator N, let $\phi'(N)$ be the result of restricting its quantifiers to the individuals (i.e. to objects not in the range of N). The truth-conditions of $\phi(N)$ are then given by the following predicative statement:

$(^{\star})\phi'(O)$ holds for some quasi-Humean operation O.

The reason the translation works is that, when there are infinitely many individuals, the cardinality of the whole domain of individuals and numbers within a minimal model will be the same as the cardinality of the individuals and so the relevant number-theoretic structure on the whole domain may be reduplicated within the subdomain of individuals.

However, no such translation can be made to work in the critical case in which the domain of individuals is finite. For suppose there were a translation taking each $\phi(N)$ into the predicative statement $\phi^{\star}$. Now in any given finite domain, we can effectively settle the truth-value of the statements $\phi^{\star}$ (since the quantifiers range over fixed finite domains of objects and concepts) and so we can effectively settle the truth-value of the statements $\phi(N)$. But this is impossible, since the statements $\phi(N)$ have the expressive power of second-order arithmetic.

This means that the predicative truth-conditions cannot be given by means of an effective translation and some other way of specifying them may be found. It might be wondered at this point why we should not simply appeal to the minimalist interpretation. For we can add an assumption to Hume's Law that has the effect of requiring that the interpretation of the number operator be minimal. The resulting theory is then categorical relative to the cardinality of the individuals: any two models with the same number of individuals are isomorphic.[24] But this then means that each statement of the theory will have a determinate predicative content as given by the range of cardinalities that the individuals may assume in any minimal model in which the statement is true.

---

[24] See Corollary III.6.7 for details.

The difficulty with this approach is to see how someone could grasp what these truth-conditions are without already having access to an infinite domain of abstract objects. To say that a statement is true, on this approach, is to say that the domain of individuals is of this size, or that size, and so on. But even if we put aside the problem of specifying the size of the domain of individuals without making reference to numbers, there remains the further problem of specifying what the relevant range of sizes should be. For the natural explanation is in terms of there being a minimal model whose domain of individuals is of that size; and this requires appeal to an infinite domain of additional objects from which the new elements in the minimal model might be drawn.

A similar difficulty besets the attempt to construct a predicative account of number on the model of the account of ordered pair given above. For how are we simultaneously to construct the number-terms and the concept-terms? One possibility is to take each concept-term to be given by a (possibly infinite) enumeration of object-terms and each number term to be given by means of a concept term. The construction should then be allowed to proceed throughout the whole set-theoretic hierarchy, since there is no natural point at which it might be stopped. But we then face an additional difficulty. For even if we are given all such terms, the truth-conditions for identity-statements will not be grounded, since they will involve quantification over all objects and hence over all number-terms. To get round this difficulty, we may restrict the quantification to the object-terms that are directly or indirectly involved in the identity statement itself. In this way, we can provide a grounded account of the truth-conditions which leads, in fact, to the minimal model.[25] However, the manner of specifying the truth-conditions is so obviously infinitistic in character—requiring, as it does, a whole set-theoretic hierarchy of expressions—that it is hard to see how we might be capable of grasping what the truth-conditions are meant to be without already presupposing an infinite ontology of abstract expressions. (Whether the account of ordered pair suffers from a similar difficulty is not so clear, since the relevant understanding of the substitutional quantifier in this case might not be thought to require grasp of an underlying ontology of expressions.)

---

[25] See the end of sect. III.7 for details.

We somehow need to provide an account of the truth-conditions that is graspable from the bottom up rather than through a sophisticated form of semantic ascent and, to this end, it will be helpful to look at the account proposed by Crispin Wright (1983: 132–45; 1998*a*: 357–68; 1998*b*: 399–405). For present purposes, I have construed Wright as attempting to show how Hume's Law might provide us with a predicative understanding of number. Although I do not think the account succeeds, as so construed, its failings will enable us to appreciate what might be required of a more adequate account. The reader should bear in mind, however, that Wright is perhaps better understood as attempting to show how Hume's Law might provide us with an *impredicative* understanding of number; and in this case, only some of the criticisms that I make will still apply.

Wright imagines a 'trainee' whose job, we are assuming, is to attain a predicative understanding of number. He is to do this in successive stages (for further details, see Wright 1983: 132–45; 1998*a*: 358–68). At each stage $n = 0, 1, \ldots,$ he introduces a new number term t_n (which intuitively will signify the number n) and then attempts to attain a logical grasp of the conditions under which the number of Fs is identical to t_n. He does this by associating a purely logical condition ϕ_n with the term t_n and deducing from Hume's Law (and what he already knows) that:

(D_n) the number of Fs $= t_n$ iff $\phi_n(F)$, for any predicate term F.

The knowledge of (D_n) is then meant to put him in a position to understand the term t_n (given a solution to the Caesar problem).

To be more specific, we take t_0 to be the term 'the number of x such that x is not self-identical' and, where $t_0, t_1, \ldots, t_{n-1}$ are the terms introduced at stages $0, \ldots, n - 1$ prior to n, we take t_n to be the term 'the number of x such that $x = t_1$ or ... or $x = t_{n-1}$' (we might call the terms $t_0, t_1, \ldots$ the *Fregean numerals*). The trainee's previous knowledge of the conditions (D_0), ..., (D_{n-1}) enables him to deduce:

$t_i \neq t_j$ for $0 \leq i < j < n$;

and his present knowledge of Hume's Law enables him to deduce:

$t_i \neq t_n$ for $0 \leq i < n$.

He is therefore in a position to deduce (D_n), where ϕ_n (F) is a logical formulation of the condition that there are exactly n Fs.[26]

I wish to direct two sets of criticisms against this proposal. For the purposes of the first, we may grant that the proposed mechanism for understanding the Fregean numerals t_0, t_1, t_2, ... will work. The criticism is then that we will still not have achieved a full understanding of arithmetical discourse. One difficulty of this sort is that it is not clear that the mechanism that Wright describes will account for anything more than the possibility, for each n = 0, 1, 2, ..., that the trainee understand the terms t_0, t_1, ..., t_n. The infinite sequence of definitions (D_0), (D_1), (D_2), ... is meant to provide him with an understanding of the Fregean numerals t_0, t_1, t_2, ... (in conjunction with a solution to the Caesar problem). First he understands t_0 with the help of (D_0); then he understands t_1 with the help of (D_1); and so on. But how is he meant to grasp this infinitary sequence of definitions? Certainly, he can grasp any finite subsequence (D_0), (D_1), ..., (D_n) of them. But how is he meant to grasp them all? Presumably he must come up with a finite compendious description of the sequence (which is what I have supplied to the reader). But this requires that he already has the concept of number! It might be thought that the fact that each of (D_0), (D_1), (D_2), ... is derivable from Hume's Law can help us here. For then our knowledge of (D_0), (D_1), (D_2), ... can be taken to be implicit in our knowledge of Hume's Law. But more than knowledge is required (if 'knowledge' is the right word, since the terms involved in the deduction are not yet understood). For from all the many statements deducible from Hume's Law, we need to be able to select a sequence of statements (D_0), (D_1), (D_2), ... that will provide us with an understanding of the terms t_0, t_1, t_2, ...; and it is the *selection* of this sequence that is providing us with so much trouble. It is unclear, in the absence of a

---

[26] We can adopt the standard first-order formulation ϕ_n (F) or the following second-order formulation:

$$\exists G \exists y (\phi_{n-1}(G) \ \& \ \neg Gy \ \& \ \forall x (Fx \leftrightarrow Gx \lor x = y)),$$

which allows for a slight simplification in the proof of (D_n) (in neither case are the complexities of Wright's (1998a: 367–8) proof at all necessary). It is worth noting that, for the purpose of finding a condition $\phi_n(F)$ for which (D_n) is deducible, it is not strictly necessary to show that the terms t_0, t_1, ..., t_{n-1} are distinct. In the case of the term t_2, for example, we might take the condition on F to be: $(t_0 = t_1 \ \& \ \phi_1(F)) \lor (t_0 \neq t_1 \ \& \ \phi_2(F))$, where ϕ_1 (F) and ϕ_2(F) are the previously defined conditions. Thus our understanding of the terms t_0, t_1, ... , need not be tied, in the way Wright suggests, to a proof of the infinitude of the number series.

prior understanding of number, how we are to grasp what this sequence is meant to be.

In any event, let us suppose that the trainee can somehow achieve an understanding of all of the Fregean numerals. He still will not possess an understanding of a term for every number. For even supposing the domain of numbers to be countably infinite, he will not yet understand a term for the number of that domain. Nor is it clear how such an understanding is to be attained. He cannot refer to it by means of an enumerative specification 'the number of x for which x = 0 or x = 1 or...', as in the case of the natural numbers, since such a specification would be infinite. Nor can he appeal to the notion of being a natural number and take the number to be that of the natural numbers. For an understanding of the definition of natural number rests upon an understanding of the locution 'the number of Fs', for variable F, which he does not yet possess. If he could somehow step outside the initial definitional process and refer to the numbers that have so far been defined, this would provide him with independent access to the notion of natural number. But the usual way of doing this already presupposes an understanding of natural number.

It might be disputed whether our trainee need have a term for every number. But having a term for every number provides him with a straightforward substitutional interpretation of quantification and, in the absence of such an interpretation, it is not clear what should take its place. It needs to be borne in mind that the interpretation of the quantifiers is not uniquely determined by Hume's Law, even when the domain of individuals is taken to be fixed. In addition to the minimalist interpretation, there are a multitude of non-minimalist interpretations; and so we need to know what it is in the trainee's understanding that determines one interpretation of the quantifiers over another.[27]

---

[27] Wright (1998a: 362–3) correctly observes that there is no separate problem over the understanding of first-order quantification though this is true, given that the numerical terms $Nx:\phi(x)$ may be formed from an arbitrary condition $\phi(x)$, for a much more straightforward reason than the one he provides: for $\exists x\phi(x)$ is equivalent, given Hume's Law, to '$Nx: \phi(x) \neq 0$'; and so first-order quantification can be eliminated in favour of the number-operator.

Given that this is so, the problem is then to show how one might understand all terms of the form $Nx:\phi(x)$. To this end, Wright appeals to the notion of rank (ibid. 363–4). But it can be shown, given his definitions, that every statement is logically equivalent to one of rank 0 (and also to one of no rank!). For it may be shown that any statement ϕ is logically

I turn to the second line of criticism, which is that Wright's trainee is not even capable of understanding the Fregean numerals. The difficulty emerges at the very first stage, when the term t_0 is introduced. For if the definition (D_0) is to be predicative at this stage, the trainee should not assume that the first-order variables range over numbers. Thus all that the definition (D_0) tells him is that if F is a *concept of individuals* then the number of Fs $= t_0$ iff there are no Fs. But this is of no help in determining whether the number of Fs is t_0 when F is not a concept of individuals and, in particular, it is of no help in determining whether t_1 (the number of objects identical to t_0) is identical to t_0, since the concept of being identical to t_0 is a concept of numbers. Even if the trainee were somehow capable of grasping that the truth of (D_0) was indifferent to the identity of the objects in the domain of quantification, he would still face difficulties in applying (D_0) to cases in which the domain was larger than that of the domain of individuals. If the domain of individuals was assumed to be finite, for example, he would be incapable of understanding what it was for the number of natural numbers to be t_0.

We might put the difficulty in the form of a dilemma: either the quantifiers in (D_0) already range over numbers, in which case the

equivalent (given that the numbers of coextensive concepts are the same) to a statement ϕ^* in which each number term is of the form Nx:Fx, for F a concept variable. But ϕ^* & Nx:Px = Nx: Px is of rank 0, when P is an unproblematic predicate (while ϕ^* itself is of no rank). Thus the appeal to rank does no work: if the trainee understands all statements of rank 0, then he effectively understands all statements whatever.

In order to take care of unranked statements, Wright (1998*b*) later suggests two ways in which his earlier account might be supplemented. One suggestion is that we might understand a second-order quantificational statement through understanding an instance (ibid. 400, case (i)). But, as we have seen, every arithmetical statement is equivalent to one in which all number-terms are of the form Nx:Fx, for variable F; and so, if this principle held, we could, by substituting $x \neq x$ for Fx, understand every arithmetical statement by understanding the term Nx: $x \neq x$ (and hence by understanding all of the statements in which that was the sole number term to occur). Clearly, within the present context, an understanding of the term Nx:x $\neq$ x cannot be taken to secure a general understanding of Nx:Fx for variable F. Another suggestion (ibid. 402) is that we might understand the term Nx:x = x through knowing that 'the condition for any particular cardinal number to be the referent of the term is that the particular numerically definite quantifier associated with that number should generate a truth when applied to 'x = x' (and similarly, I assume, for any predicate Fx in place of 'x = x', so that this could be the general account of our understanding of number terms). But this is doubly circular, since we can have no general understanding of what numerical quantifier should be associated with a given term without already having a general understanding of what the term refers to and since the quantifier associated with a given term must be understood impredicatively as already applying to numbers as well as to individuals.

definition is impredicative; or they only range over individuals, in which case the definition is inadequate.

There is a further difficulty with the definitions (D_0), (D_1), ..., which concerns the left-rather than the right-hand side. Consider again the case of (D_0). We are attempting to understand the term t_0 and, to this end, we provide the partial definition:

(D_0) the number of Fs = t_0 iff there are no Fs.

But does not our understanding of t_0 then depend upon a prior understanding of the number operator?[28] Suppose that the operator 'the number of' was so understood that, in case there were no Fs, the number of Fs was taken to be Caesar. (D_0) would then provide us with an understanding of the numeral t_0 for which it was identical to Caesar. But this then means that we must already presuppose the required understanding of the number operator, if the definitions (D_0), (D_1), ... are to yield the required understanding of the number terms.

It might be thought that this was a general difficulty with all forms of abstraction principle. But this is not so. Consider the principle for character:

the character of person P = the character of person Q iff P is of-the-same-character as Q.

Our understanding of the term 'the character of Judas' need not be taken to depend upon a prior understanding of the character operator, since we can think of the above principle as *simultaneously* determining the reference of the character terms and the character operator. Such a simultaneous understanding does not appear to be possible in the case of the number; and it is only when we attempt to replace it with a successive understanding of various number terms that the present difficulty forces itself upon us.

Let me conclude with a general complaint about the present approach. In all the cases so far considered, we are somehow meant to be able to read off from certain principles how the notion of number is to be understood and, even though the principles are not themselves predicative in form, it is somehow meant to be apparent how they provide us with a predicative understanding.

---

[28] The term t_0 is also of the form 'the number of Fs'. But the present difficulty arises even if, as is perhaps desirable, the term t_0 is taken to be a constant.

This is all intolerably ad hoc and a modicum of logical rigour would appear to demand, first, that it be manifest from the form of the definition how it is to be understood and, second, that it should be possible to provide a definition that is explicitly predicative in form whenever we think that a predicative understanding can indeed be attained. Without such rigour, we cannot pretend to have any real understanding either of how these definitions are meant to work or of how they should be presented.

We have seen that there are three main problems with CP: achieving a satisfactory solution to the Caesar problem, one applicable to all forms of contextual definition; showing how referential determinacy may be achieved, without collapsing contextual definition into a standard form of implicit or explicit definition; and providing a predicative account of number and the like. Although these difficulties may seem severe, perhaps insurmountable, I believe that they may be overcome by adopting the procedural form of postulationism mentioned in the Preface. The basic idea behind this alternative approach is that, instead of stipulating that certain statements are to be true, one specifies certain procedures for extending the domain to one in which the statements will in fact be true. These procedures can be stated without invoking an abstract ontology; they achieve referential determinacy; and their legitimacy does not depend upon the prior knowledge that the objects which are to be introduced into the domain already exist.

III

The Analysis of Acceptability

TH E present part centres on several questions concerning the acceptability of a theory of abstraction. Sections 1–3 are devoted to preliminaries: sect. 1 describes the basic languages and systems of interest to us; sect. 2 specifies the semantics for the languages; and sect. 3 establishes some basic proof- and model-theoretic results. Sections 4–6 deal with various requirements for acceptability: sect. 4 contains some preliminary results on tenability, which will later be refined in the section on invariance; sect. 5 introduces the notion of a generated model; and sect. 6 shows how an abstraction theory may be categorical with respect to its generated models. Sections 7–9 are devoted to the topic of invariance: sect. 7 provides an intrinsic analysis of those criteria of identity that are invariant and non-inflationary; sect. 8 builds on these results to determine when the class of all invariant methods of abstraction will not 'hyperflate' over a given domain; and sect. 9 shows how the results on invariance may be internalized to a theory of abstraction and limits thereby set on which abstraction principles can consistently be assumed.

1. Language and Logic

The underlying language of our investigations is the language L^2 of second-order logic. This is obtained from the usual language of first-order logic by adding variables for relations of arbitrary finite degree. The relational variables apply to the first-order objectual terms in the same way as the relational constants; and the quantifiers apply to both the first- and the second-order variables. We suppose that the language contains a symbol for identity but only allow it to flank terms of first order.

We shall occasionally appeal to the language L^3 of third-order logic. This is obtained from L^2 by adding relational variables, and possibly also some relational constants, of arbitrary third-order degree. In particular, it will contain a binary relational variable of third order whose first term is of second order and whose second term is of first order.

Both the languages L^2 and L^3 may have non-logical constants where they have variables; and so, strictly, speaking, there are different languages for different choices of constants. But our main interest is in languages without non-logical constants. These languages and their associated formulas will be called *(purely) logical.*

For the most part, our notation will either be standard or intelligible from the context. We reserve: C, D, E, etc. for unary relational symbols; R, S, T, etc. for the binary ones; and P, Q, P', Q', etc. for those of arbitrary degree. Relational constants and variables of third order will be distinguished by being displayed in boldface. We follow Frege in thinking of the first-order terms as standing for *objects* and the unary second-order terms as standing for *concepts.*

The language $L^\S$ of *abstraction* is obtained from L^2 by adding an operator $\S$ of second-order abstraction. This applies to a concept-term C and results in an object-term $\S$C. Where $\mathbf{\S}$ (boldface) is a set of abstraction operators, we may also form the result $L^\S$ of adding all of the members $\S$ of $\mathbf{\S}$ to L^2. We should perhaps have treated $\S$ as a variable-binding device that applies to a variable x and a formula ϕ to form an object-term $\S x\phi$; what we now express as $\S$C would then be expressed as $\S x Cx$. Whatever the philosophical reasons for preferring the one notation to the other, nothing of great technical significance hangs on the choice; and it will turn out that the variable-free notation is more convenient for our purposes.

The following abbreviations within the language L^2 will be useful (the type of the various terms occurring in the definienda being evident from the context):

Domain: $x \in Dm$ (R) for $\exists y Rxy$;
Range: $x \in Rg$ (R) for $\exists y Ryx$;
Field: $x \in Fld(P)$ for $\exists x_2 \ldots x_n$ ($Pxx_2x_3 \ldots x_n \lor Px_2xx_3 \ldots x_n \lor Px_2x_3 \ldots x_nx$);
Reflexivity: Refl (R) for $\forall x Rxx$;
Symmetry: Sym (R) for $\forall xy$ (Rxy $\rightarrow$ Ryx);
Transitivity: Trans (R) for $\forall xyz$ (Rxy & Ryz $\rightarrow$ Rxz);
Equivalence: Eq(R) for Refl (R) & Sym (R) & Trans (R);
Inclusion: $P \subseteq Q$ for $\forall x_1 \ldots x_n$ ($Px_1 \ldots x_n \rightarrow Qx_1 \ldots x_n$);
Coextensiveness: $C \equiv D$ for $\forall x$ (Cx $\leftrightarrow$ Dx);

Complementation: D compl C for $\forall x(Dx \leftrightarrow \neg Cx)$;

Mapping: $P \rightarrow_R Q$ for $\forall x_1 \ldots x_n \forall y_1 \ldots y_n (\bigwedge Rx_i y_i \rightarrow (Px_1 \ldots x_n \leftrightarrow Qy_1 \ldots y_n)$;

One-to-one: $1\text{--}1(R)$ for $\forall xyz(((Rxy \ \& \ Rxz) \lor (Ryx \ \& \ Rzx)) \rightarrow y = z)$.

Equinumerosity via R: $C \ eq_R \ D$ for $C \rightarrow_R D \ \& \ 1\text{--}1(R) \ \&$ $\forall x((x \in Dm(R) \leftrightarrow Cx) \ \& \ (x \in Rg(R) \leftrightarrow Dx))$;

Equinumerosity: $C \ eq \ D$ for $\exists R(C \ eq_R \ D)$;

Inequality: $C \leq D$ for $\exists D'(D' \subseteq D \ \& \ C \ eq \ D')$;

Biequinumerosity; $C \ beq \ D$ for $C \ eq \ D \ \& \ \forall C', \ D' \ [(C' \ compl \ C) \ \& \ (D' \ compl \ D)] \rightarrow C' \ eq \ D')$;

Permutation: Perm (R) for $1\text{--}1(R) \ \& \ \forall x(x \in Dm(R) \ \& \ x \in Rg(R))$.

The logic L^2 of second order is based on the second-order language L^2. (So, strictly speaking, we obtain different logics for different choices of the non-logical constants.) The logic contains the usual truth-functional and quantificational principles for first- and second-order terms. However, I assume that these principles are formulated so as to be tolerant of an empty domain. The logic L^2 contains, in addition, the *Comprehension Scheme* for relations of arbitrary degree: where P is a relational variable of degree n which does not occur free in ϕ, $\exists P \forall x_1 \ldots x_n \ (Px_1 \ldots x_n \leftrightarrow \phi)$ is to be an axiom.

L^3 is the language of the third-order logic L^3, which is obtained from L^2 by extending the truth-functional and quantificational principles in the obvious manner and by adding a comprehension scheme for third-order relations. For example, in the case of a relation between a concept and an object, it would contain axioms of the form $\exists F \forall C \forall x(F(C, x) \leftrightarrow \phi)$.

$L^\S$ is the language for the various theories of abstraction. Let ϕ be a formula whose only free variables are the concept variables C and D. Then Φ is the formula $\S C = \S D \leftrightarrow \phi \ (C, D)$; and the abstraction theory T^Φ is the result of adding Φ as an axiom to L^2 (as defined over the language that results from adding $\S$ to the language of ϕ). Similarly, when $L^\S$ is a language containing several abstraction operators and ϕ associates a formula $\phi_\S$, meeting the specifications above, with each $\S$ in $\S$, the *generalized* abstraction theory T^Φ is the result of adding all of the formulas $\Phi_\S$ as axioms to the logic L^2 (based on $L^\S$).

The formula ϕ is called an *identity criterion* since it provides a criterion for two abstracts to be the same, and the function ϕ is called a *system* of identity criteria. The associated formula Φ is called *Abstraction* or the *Abstraction Principle* (*for* ϕ).

The identity criterion ϕ (the associated principle Φ, the theory T^ϕ, and the operator § in T^ϕ) are said to be *logical* if ϕ contains no occurrence of non-logical constants or of §; and they are said to be *grounded* if ϕ contains no occurrence of §. Logical criteria are also called *L-criteria*. Within a logical language, logical and grounded criteria will, of course, coincide.

These definitions can be extended to generalized theories T^ϕ and even to theories in which the identity conditions are permitted to be infinitary. Let us say that § *precedes* §′ in the generalized theory T^ϕ— in symbols, § < §′—if § occurs in $\phi_{§'}$. Then T^ϕ is said to be *logical* if all of its identity conditions $\phi_§$ are logical; and it is said to be *grounded* if the associated relation < of precedence is well founded, i.e. if there is no infinite sequence $§_1, §_2, \ldots$ of abstraction operators and of corresponding identity conditions $\phi_1, \phi_2, \ldots$ with the property that $§_{i+1}$ occurs in ϕ_i for $i = 1, 2, \ldots$. Regarding abstraction principles as forms of definition, the requirement that a system of such principles be grounded can be seen as part of the general requirement that definitions be grounded.

For any well-founded relation < on the operators in §, there exists a compatible well-ordered sequence $< §_\xi: \xi < \alpha >$ of the members of §, i.e. one with the property that $§_\nu < §_\xi$ only if $\nu < \xi$ for any ordinals $\nu, \xi < \alpha$. Accordingly, let us say that a generalized theory T^ϕ is *definable over* a well-ordered sequence $< §_\xi: \xi < \alpha >$ of its operators §, or is a $< §_\xi: \xi < \alpha >$-theory, if the well-ordering of the operators § is compatible with the relation of precedence for T^ϕ. Thus in any $< §_\xi: \xi < \alpha >$-theory T^ϕ, $\phi_§$, for § = $§_\xi$, will only contain operators $§_\nu$ for $\nu < \xi$.

The following abbreviations within $L^§$ will be useful:

Abstract: Ab (x) for $\exists C(x = §C)$;
Individual: I (x) for $\neg Ab(x)$;
Null abstract: NAb (x) for $\exists C(\forall x \neg Cx \ \& \ x = §C)$;
Universal abstract: UAb (x) for $\exists C(\forall xCx \ \& \ x = §C)$.

We shall also be interested in *restricted* abstraction theories. Let ϕ be an identity criterion, as before, and let $\psi = \psi(C)$ be a formula whose sole free variable is C. Then the *restricted abstraction principle* Φ_ψ is the formula $\psi(C) \ \& \ \psi(D) \rightarrow (§C = §D \leftrightarrow \phi(C, D))$ and the *restricted abstraction theory* T^ϕ, ψ is the result of adding Φ_ψ to the logic L^2 based on the language of Φ_ψ. Thus in a restricted principle, the application of the identity criterion is limited to those concepts that conform to a certain constraint.

We use $\vdash^2$, $\vdash^3$, $\vdash^\phi$ and $\vdash^\phi$ to indicate provablity in L^2, L^3, T^ϕ and T^ϕ, respectively.

2. Models

We outline the model theory for the various languages and define some basic model-theoretic notions.

A *model M for* the language L^2 is a quadruple (M, R, e, v), where M (the domain of *objects*) is a (possibly empty) set, R is a function that takes each n, for n $\geq$ 0, into a set (the domain of n-ary *relations*), e (*extension*) is a function that takes each member of R_n into a subset of M^n (the set of n-tples from M), and v (*valuation*) is a function that takes each object constant into a member of M and each n-ary relation constant into a member of R_n.

A *model $M^\S$ for* $L^\S$ (also called a *§-model*) is a quintuple $(M, R, e, \S, v)$, with $(M, R, e\ v)$ as before and with $\S$ (the *abstractor*) a function from the domain R_1 of concepts into the domain M of objects. The model $M^\S = (M, R, e, \S, v)$ is said, in such a case, to be a *§-expansion of* the model $M = (M, R, e, v)$. Similar definitions can be given for the other languages. But, of course, when the language contains several abstraction operators **§**, the model will contain an abstractor § for each of the operators. If *M* is a **§**-model then the *reduction $M_\S$ of M*, for § ϵ **§**, is the §-model with the one abstractor § in place of the many abstractors in *M*.

Given a model $M^\S = (M, R, e, \S, v)$, the objects within the range of § are called *abstracta* and the remaining objects of *M* are called *concreta* or *individuals*. Similarly, in a model with several abstractors, the abstracta are the objects lying in their range and the individuals are the remaining objects. Note that individuals should only be taken to be individuals in a relative sense; even if they cannot be obtained as a result of the methods of abstraction represented within the model, they may still result, intuitively speaking, from some other kind of abstraction.

We use *I* for the set of individuals and *A* for the set of abstracta; and we may employ an index *M*, both here and elsewhere, to indicate dependence upon the underlying model. The model *M* is said to be *pure* when its domain *I* of individuals is empty.

The model $M = (M, R, e, v)$ (or its expansion $M^\S$) is said to be *extensional* if extensionally equivalent relations are the same, i.e. if $e(R) = e(S)$ implies $R = S$ for any R and S in $R_n, n > 0$. A subset of $M^n, n > 0$, is said to be *represented in M* if it is the extension of some relation in R_n; and the model M (or $M^\S$) is said to be *full* if every subset of M^n, for $n = 1, 2, \ldots$, is represented.

The model M (or $M^\S$) is said to be *set-theoretic* if M consists entirely of urelements (i.e. non-sets) and e is an identity function on $\bigcup R_n$; and M (or $M^\S$) is said to be *standard* if it is both set-theoretic and full. Clearly, any set-theoretic model is extensional. We shall assume, for convenience, that there is a proper class U of urelements and that all of the urelements that we need are drawn from U.

Any standard model M of a language L^2 without non-logical constants is uniquely determined by its domain M and hence is uniquely determined up to isomorphism by the cardinality c of its domain. We denote this model (as given up to isomorphism) by M_c.

Three kinds of abstractor will be of special importance in what follows. These are the *cardinal* abstractors $\S$, which give the value $\S(C) = \text{card}(C)$ for each subset C of the domain M, the *bicardinal* abstractors, which give the value $\S(C) = <\text{card}(C), \text{card}(M-C)>$ for each subset C of M, and the *divisor* abstractors, those with a range of cardinality ≤ 2.

Let M be a model for $L^\S$ (or for $L^\S$). Then with each abstractor $\S$ in M we may associate an equivalence relation $\equiv_\S$ on concepts defined by: $C \equiv_\S D$ iff $\S(C) = \S(D)$. The associated equivalence classes $|C|_\S$ and partition $P_\S$ may then be defined in the usual way.

We may distinguish three ways in which a model may be separated. The model M for $L^\S$ is said to be *strictly separated* if the ranges of the different $\S_i$s are disjoint. No two methods of abstraction yield the same abstract. The model M for $L^\S$ is said to be *separated* (*simpliciter*) if, for any abstractors $\S$ and $\S'$ in M, $\S(C) = \S'(D)$ holds iff $|C|_\S = |D|_{\S'}$. Two abstracts, of different types, are identified iff their associated equivalence classes are the same. The model M for $L^\S$ is said to be *weakly separated* if, for any abstractors $\S$ and $\S'$ in M, $\S(C) = \S'(D)$ holds only if $|C|_\S = |D|_{\S'}$. Two abstracts, of different types, are the same if and only if their associated equivalence classes are the same.

Satisfaction and truth for the models of our various languages are defined in the usual way. In particular, $\S C$ will denote $\S(C)$, where C is the concept from R_1 assigned to the variable C. With truth as so defined, any model will be a model of the theorems of second-order logic without Comprehension. From henceforth, we shall take our models also to be models for Comprehension unless there is an explicit rider to the contrary.

When M is a model for one of our languages L and ϕ is a formula of L with free variables $x_1, \ldots, x_m, P_1, \ldots, P_n$ (in that order of appearance), we take the *extension* $E_{\phi, M}$ *of* ϕ *in* M to be the set $\{< x_1, \ldots, x_m, P_1, \ldots, P_n > : M \models \phi[x_1, \ldots, x_m, P_1, \ldots, P_n]\}$. (The subscript M may be omitted from $E_{\phi, M}$, and elsewhere, when it is obvious from the context).

We are especially interested in the case in which $\phi = \phi(C, D)$ is an identity criterion (with free variables C and D). In this case, $E_{\phi, M}$ is a relation on concepts, i.e. a subset of $R_1 \times R_1$, and is called the *criterial relation in M*. Should $E_{\phi, M}$ be an equivalence relation, it will induce a partition $P_{\phi, M}$ over R_1 in the obvious way; and when M is a standard model, $E_{\phi, M}$ and $P_{\phi, M}$ will be an equivalence and a partition over the subsets $\wp(M)$ of M.

Given a set M of urelements, we take a *(local) set relation over M* to be a subset of $\wp(M) \times \wp(M)$. A *global set relation* is then taken to be a (proper class) function that takes each set M of urelements into a local set relation over M. Each logical identity criterion ϕ will determine a global set relation, also denoted by E^{ϕ}, which takes each set M of urelements into $E_{\phi, M}$, where M is the standard model with domain M.

3. Preliminary Results

We establish various elementary model- and proof-theoretic results for our systems. We begin with the proof theory.

Where $\phi(C)$ is a a formula containing free occurrences of the concept variable C, let $\phi(D)$ be the result of replacing any free occurrence of C in $\phi(C)$ with a free occurrence of D. Then within second-order logic, we may show:

Lemma 1 (Extensionality). $\vdash^2$ C $\equiv$ D $\rightarrow$ $(\phi(C) \leftrightarrow \phi(D))$.

Proof. By a straightforward induction on the complexity of the formula $\phi(C)$.

This result fails to hold in third-order logic or if we allow identity on concepts. For in the first case, C $\equiv$ D $\rightarrow$ $(F(C) \leftrightarrow F(D))$ is unprovable for F a third-order concept term; and in the second case, $C \equiv D \rightarrow (C = C) \leftrightarrow (C = D)$ is unprovable. The result also fails to hold in an arbitrary abstraction theory. For let the theory be the degenerate one T^{ϕ} in which ϕ is the formula $\S C = \S D$. Then $C \equiv D \rightarrow \S C = \S D$ is unprovable.

We may establish the fixed-point theorem within the confines of second-order logic. For $\phi(C, D)$ a formula (which we treat as defining a function with argument C and value D), we adopt the following abbreviations:

> Functionality: Func (ϕ) for $\forall C \exists D(\phi(C, D)$ & $\forall E(\phi(C, E) \rightarrow E \equiv D))$;
> Monotonicity: Mon (ϕ) for $\forall C, D, C', D'(\phi(C, D)$ & $\phi(C', D')$ & $C \subseteq C' \rightarrow D \subseteq D')$;
> Fixed Point over B: $FP_{\phi, B}(F)$ for $B \subseteq F$ & $\phi(F, F)$;
> Least Fixed Point over B: $LFP_{\phi, B}(F)$ for $FP_{\phi, B}(F)$ & $\forall E(FP_{\phi, B}(E) \rightarrow F \subseteq E)$.

The fixed-point theorem now takes the following form:

Theorem 2 (Least Fixed Point). $\vdash^2$ Func(ϕ) & Mon(ϕ) $\rightarrow \exists F$ $(LFP_{\phi, B}(F))$.

Proof. We present an informal proof, but in a way that makes clear how the proof is to be formalized. We first consider the case in which B is empty. Define a concept C to be *determining* if $\forall D(\phi(C, D) \rightarrow D \subseteq C)$. Now say that an object is *determined* if it falls under every determining concept; and let F be the concept of being a determined object. We may then readily show that F is a 'least fixed point'.

In the general case, let ϕ^+ be the formula: $\exists C^+[\forall x(C^+ x \leftrightarrow (Bx \lor Cx))$ & $\phi(C^+, D)]$. Applying the first case to this formula then yields the desired result.

We turn now to abstraction theories and show first that the relation on concepts determined by the identity criterion $\phi(C, D)$ is an equivalence:

Lemma 3 (Equivalence). In the abstraction theory T^{ϕ}, the formulas $\phi(C, C)$, $\phi(C, D) \rightarrow \phi(D, C)$ and $[\phi(C, D)$ & $\phi(D, E)]$ $\rightarrow \phi(C, E)$ are all theorems.

Proof. Using the abstraction principle, the above formulas 'translate' into theorems of the first-order logic of identity. For example, $\phi(C, C)$ translates into $\S C = \S C$.

Note that the above proof makes no appeal to the Comprehension Scheme in T^ϕ.

Using the above lemma, we can show that coextensive concepts must be identified in any abstraction theory T^ϕ in which ϕ is grounded (i.e. §- free):

Lemma 4 (Inclusion). $\vdash^\phi$ $C \equiv D \to \phi(C, D)$ for grounded ϕ.

Proof. Given that ϕ is §-free, it follows by the extensionality lemma that $\vdash C \equiv D \to (\phi(C, C) \to \phi(C, D))$. By the equivalence lemma, $\vdash \phi(C, C)$. Hence $\vdash C \equiv D \to \phi(C, D)$.

The extensionality lemma can now be extended to abstraction theories:

Lemma 5 (Extended Extensionality). $\vdash^\phi$ $C \equiv D \to (\psi(C) \leftrightarrow \psi(D))$ for grounded ϕ and any ψ.

Proof. As before by induction. But to take care of the presence of § in the language, we need to show that $C \equiv D \to §C = §D$ is a theorem. But $C \equiv D \to \phi(C, D)$ is a theorem by the inclusion lemma; and $\phi(C, D) \to §C = §D$ follows from the abstraction principle ϕ.

The inclusion and extensionality lemmas can be extended by means of the obvious transfinite induction to generalized abstraction theories that are grounded.

Invariance under permutations will play a major role in our attempt to understand what can be conveyed by a logical criterion of identity. It captures part of what Frege might have meant by the generality of logic in sect. 3 of the *Grundlagen*. We have, in the first place, the following syntactic formulation:

Lemma 6 (Permutation). Suppose that $\psi = \psi(x_1, \ldots, x_m, P_1, \ldots, P_n)$ is a logical formula whose only free variables are the distinct variables $x_1, \ldots, x_m, P_1, \ldots, P_n$. Let $Y_1, \ldots, Y_m$ and $Q_1, \ldots, Q_n$ be distinct variables which do not occur in ψ, with each Q_j of the same arity as P_j. Then $\vdash^2$ (Perm(R) & $\bigwedge_i Rx_iy_i$ & $\bigwedge_j P_j \to_R Q_j) \to [\psi(x_1, \ldots, x_m, P_1, \ldots, P_n) \to \psi(y_1, \ldots, y_m, Q_1, \ldots, Q_n)]$.

Proof. By induction on the complexity of the formula. For the case that ψ is a second-order quantification we need to appeal to comprehension.

As a special case of the above result, we have:

Corollary 7 (Invariance). Suppose that $\phi = \phi(C, D)$ is a logical identity criterion. Then $\vdash^2$ (Perm(R) & $C \to_R C'$ & $D \to_R D'$) $\to$ ($\phi(C, D) \to \phi(C', D')$).

Thus we may prove within second-order logic that logical identity criteria are invariant under permutations of the universe of objects: the images of concepts related by a criterion are also related by the criterion. Note that the lemma may fail when the formula $\psi(x_1, \ldots, x_m, P_1, \ldots, P_n)$ contains non-logical constants or the operator §. For from the fact that Perm(R) & $C \to_R D$ we cannot infer $Ca \to Da$ for a an object constant (and similarly for relation constants); and from the fact that Perm(R) & $C \to_R D$ we cannot infer §$C =$ §D. On the other hand, we should note that a version of this result will continue to hold when the language is enriched with further logical constants, be they finitary or infinitary in character.

We now consider the model-theoretic results. The first tells us that Abstraction does nothing to determine the identity of abstracts (beyond their being objects). Let us note, for the purposes of the proof, that an §-model M is a model for the abstraction principle ϕ (corresponding to the criterion ϕ) just in case for every $C, D \in R_1$, $\S(C) = \S(D)$ iff $<C, D> \in E_{\phi, M}$.

Lemma 8 (Switching). Suppose that M is a model for grounded ϕ and that f is a one-to-one function from M into M. Let M' be the result of replacing the abstractor § in M with the compositional map $\S' = f \circ \S$. Then M' is also a model for ϕ.

Proof. M is a model for ϕ iff for every $C, D \in R^1$, $\S(C) = \S(D)$ iff $<C, D> \in E_{\phi, M}$; while M' is a model for ϕ iff for every $C, D \in R^1$, $\S'(C) = \S'(D)$ iff $<C, D> \in E_{\phi, M'}$. But given that f is one-to-one, $\S(C) = \S(D)$ iff $\S'(C) = \S'(D)$; and given that ϕ is §-free, $E_{\phi, M} = E_{\phi, M'}$.

This proof is related to Frege's argument in sect. 10 of the *Grundgesetze*. (See Schroeder-Heister (1987) and T. Parsons (1987) for a careful discussion of his argument.) Our proof fails in case ϕ is ungrounded. For example, let ϕ be the principle §$C =$ §$D \leftrightarrow \exists x$ (UAb(x) & $Cx \leftrightarrow Dx$), in which concepts are identified when they agree on the universal abstract; and let M be a model for ϕ with two elements. Switching the universal abstract with any other object then gives a structure M' which is not a model for ϕ.

Call a model M for $L^\S$ *§-extensional* if $\S(C) = \S(D)$ whenever $C, D \in \mathbf{R}_1$ and $e(C) = e(D)$. Thus a §-extensional model is one which conforms to the *extensionality principle*: $C \equiv D \rightarrow \S C = \S D$. With any §-extensional model $M = (M, \mathbf{R}, e, \S, v)$ whose domain M consists of urelements, we may associate a set-theoretic model $M^s = (M, \mathbf{R}', e', \S', v')$ by setting $\mathbf{R}'_n = \{e(R): R \in \mathbf{R}_n\}$, $e'(R) = R$ for $R \in \mathbf{R}'_n$, $\S'(e(C)) = \S(C)$ for $e(C) \in \mathbf{R}'_1$, $v'(\mathbf{R}) = e(v(\mathbf{R}))$ for R a relational constant, and $v'(a) = v(a)$ for a an object constant. In effect, extensionally equivalent relations are identified with their extensions. It is then readily shown:

Lemma 9. Suppose that M is an §-extensional model $(M, \mathbf{R}, e, \S, v)$ with associated set-theoretic model M^s. Then $M \models \phi[P_1, \ldots, P_m, x_1, \ldots, x_n]$ iff $M^S \models \phi[e(P_1), \ldots, e(P_m), x_1, \ldots, x_n]$.

This lemma will justify us in confining our attention to the set-theoretic models of an abstraction theory T_ϕ in case ϕ is §-free.

We now provide a model-theoretic account of invariance under permutation. Say that a global-set relation $\mathbf{R}$ is *invariant* if, for each non-empty subset M of urelements, one-to-one map f from M onto N, and subsets C and D of M, $\mathbf{R}_M(C, D)$ implies $\mathbf{R}_N(f(C), f(D))$; and say that a local set relation R over M is *invariant* if, for any subsets C and D of M and permutation f on M, $R(C, D)$ implies $R(f(C), f(D))$. Clearly, if the global set relation $\mathbf{R}$ is invariant then so are each of the local set relations $\mathbf{R}_M$.

Lemma 10 (Outer Invariance). For ϕ a logical criterion, the corresponding global set relation E_ϕ is invariant.

Proof. Suppose that M and N are non-empty sets of urelements, C and D are subsets of M, and f is a one-to-one map from M onto N. Let M and N be standard models (for a purely logical language) with respective domains M and N. Suppose that $\mathbf{R}_M(C, D)$ for $\mathbf{R}$ the global relation E_ϕ associated with ϕ. Then $M \models \phi[C, D]$. Now f can be extended to an isomorphism from M onto N; and so, $N \models \phi[f(C), f(D)]$. But then $\mathbf{R}_N(f(C), f(D))$.

We finally consider those identity criteria whose behaviour is independent of the underlying domain. Such criteria play an important role in the study of generated abstraction models. We provide a syntactic and semantic criterion for such domain independence and connect the two. Given the sequence S of relational and objectual terms $P_1, \ldots, P_m, t_1, \ldots, t_n$, say that the formula ϕ is *relativized to* S if each (universally) bound objectual variable x is relativized to the formula $x \in \mathrm{Fld}(P_1) \vee \ldots \vee x \in \mathrm{Fld}(P_m) \vee x = t_1 \vee \ldots \vee x = t_n$ and each (universally) bound relational variable P is relativized to the formula $\forall x(x \in \mathrm{Fld}(P) \rightarrow x \in \mathrm{Fld}(P_1) \vee \ldots \vee x \in \mathrm{Fld}(P_m) \vee x = t_1 \vee \ldots \vee x = t_n)$. When S is the sequence of all of the objectual and relational terms that occur—either as free variables or as non-logical constants—in the very formula ϕ, we say that ϕ is *restricted*.

On the semantic side, let $M = (M, R, e, v)$ be a second-order model and N a subset represented in M. We let the *restriction* or *submodel* $N = M/N$ of M induced by N be the model (N, R', e', v), where $R'_n = \{P \in R_n : e(P) \subseteq N^n\}$, and e' is the restriction of e to $\bigcup_n R'_n$. For a §-model $M^\S = (M, R, e, \S, v)$, we also require that $\S'$ in the restriction $M^\S/N$ should be the restriction of $\S$ in M to R'_1. The values of the non-logical constants must be defined in N for the restriction to be well defined. However, we shall allow $\S'(C)$ to be undefined when $\S(C)$ does not belong to N but C belongs to R'_1. In this case the restriction $M^\S/N$ will be said to be a *partial* model, i.e. one whose abstraction operation is not always defined; and otherwise it will be said to be *total*. For the most part, our interest is in standard models of L^2; and, in this case, every subset N of the domain M of the model M will determine a restriction of M.

It is readily shown that the restriction N verifies the comprehension principle $\exists P \forall x_1 \ldots x_n (P x_1 \ldots x_n \leftrightarrow \phi)$ given that M does. For $M \models \exists P \forall x_1 \ldots x_n (P x_1 \ldots x_n \leftrightarrow \phi^C)$, where ϕ^C, for C a new concept variable, is in the obvious sense the relativization of ϕ to C. Letting C 'be' a concept whose extension is N then yields in M a concept 'P' which, in N, will correspond to ϕ.

Let $\phi = \phi(P_1, \ldots, P_m, x_1, \ldots, x_n)$ be a formula of L^2 whose free variables are as displayed. Then ϕ is said to be *absolute* if for any model M for the language of ϕ and restriction N of M, $M \models \phi[P_1, \ldots, P_m, x_1, \ldots, x_n]$ iff $N \models \phi[P_1, \ldots, P_m, x_1, \ldots, x_n]$ for any relations $P_1, \ldots, P_m$ (of appropriate type) and objects $x_1, \ldots, x_n$ from N. Thus whether the formula holds of certain objects and relations is unaffected by what in the model lies beyond those objects and the extensions of the relations.

Suppose that M is a model for a language L^2 with finitely many non-logical constants; and suppose that it verifies Comprehension. Let $P_1, \ldots, P_m, x_1, \ldots, x_n$ be relations and objects from M.

There will then be a smallest restriction N of M to contain $P_1, \ldots, P_m, x_1, \ldots, x_n$. If there are no non-logical constants in the language, the domain of N will be the union of the fields of $e(P_1), \ldots, e(P_m)$ with $\{x_1, \ldots, x_n\}$; while if there are non-logical constants in the language, the objects from their values must also be included. In each case, the domain N will be definable in M by an appropriate disjunction; and it should be clear that a formula $\phi = \phi(P_1, \ldots, P_m, x_1, \ldots, x_n)$ will be absolute just in case satisfaction in M is equivalent to satisfaction in the smallest restriction of M for each choice of relations and objects $P_1, \ldots, P_m, x_1, \ldots, x_n$.

It may be shown by a straightforward induction:

Lemma 11 (Absoluteness). Any restricted second-order formula is absolute.

Clearly, the result also holds for any formula that is provably equivalent to a restricted second-order formula. For example: the formula $C \equiv D$ is equivalent to the restricted formula $\forall x \, (Cx \vee Dx \to Cx \leftrightarrow Dx)$ and is therefore absolute. Similarly, for the formula C eq D. On the other hand, the formula C beq D is not absolute and therefore is not provably equivalent to any restricted formula.

Absolute identity criteria define an especially simple form of global relation. Say that a global relation R is *internal* if, for any subsets X and Y of urelements and subsets C and D of $X \cap Y$, $R_x(C, D)$ implies $R_Y(C, D)$. Equivalently, R will be internal if $R_{C \cup D}(C, D)$ holds iff $R_Y(C, D)$ holds for any $Y \supseteq C \cup D$.

Corresponding to any global set relation R is an ordinary (unindexed) relation R that holds between any two sets C and D just in case $R_X(C, D)$ holds for some set X of urelements. In general, a global relation cannot be recovered from the corresponding unindexed relation. But in case the global relation is internal, it can be: for $R_X(C, D)$ holds, for $C, D \subseteq X$, just in case $R(C, D)$ holds.

We say that the global relation R is *internally invariant* (or *I-invariant* for short) if it is both internal and invariant; and we say that a local relation R over M is *internally invariant* (*I-invariant*) if $R(C, D)$ implies $R(f(C), f(D))$ for any subsets C and D of M and any one-to-one function f from $C \cup D$ into M. Clearly, when R is internally invariant, then so are each of the corresponding local relations R_M. It is readily shown from Lemmas 10 and 11 that:

Lemma 12. For ϕ a restricted L-criterion, E_ϕ is internally invariant.

4. Tenability

We consider various formal conditions that might be proposed for the truth or correctness of a principle of abstraction. The philosophical significance of the conditions has been discussed towards the end of sect. I.1; and further conditions are considered in some of the later sections.

Given a §-free model M, we say that the identity criterion ϕ (or the associated abstraction principle Φ) is *tenable on M* if some expansion of M is a model of Φ and, for d a cardinal, we say that ϕ is *d-tenable on M* if some expansion $M^\S$ of M is a model of Φ and contains exactly d individuals. When the model M is standard and the criterion ϕ is logical, all that matters about M is its cardinality. We may therefore say that the L-criterion ϕ is *tenable on the cardinal c* if it is tenable on a standard model M with domain of cardinality c and that it is *d-tenable on c* if it is tenable on a standard model M whose domain is of cardinality c and whose subdomain of individuals is of cardinality d.

The above condition is relative to a model or a cardinal. We may obtain an absolute condition by declaring an L-criterion ϕ to be *stable* if, for some cardinal c, ϕ is tenable on d for any cardinal $d \geq c$. I have here presupposed that the underlying model is full; but the account could readily be accommodated to other, more restrictive, conceptions of what concepts and relations are admitted into a model. If ϕ is stable, it will be said to *stabilize* on the smallest cardinal c for which ϕ is tenable on every $d \geq c$.

Other absolute notions might also be distinguished. For example, we might say that an L-criterion ϕ is *generally* tenable if it is tenable on each infinite cardinal, or that ϕ is *indefinitely* tenable if there is no greatest cardinal upon which it is tenable.

It might be thought that the requirement on cardinality in the definition of stability is too weak. For given a model whose domain of objects is of transfinite cardinality c, one might want to insist that the domain of individuals (relative to the given abstractor) should be capable of being of any reasonable size $d \leq c$. However, it may be argued that this further requirement is automatically met. For let us suppose that the cardinality of the abstracts is $c' \leq c$. If $c' = c$, we may—by means of the switching Lemma 3.8 above—allow d to be of any cardinality $\leq c$. On the other hand, if $c' < c$, it may plausibly be maintained that d should be equal to c. For there will be at least as many abstracts of some sort or another as individuals (in the absolute sense) and so if there are only $c' < c$ abstracts of the given sort there must be at least c abstracts of another sort, which will, in the required relative sense, be individuals. But given $d = c$, we may again by means of the switching lemma allow d to be the cardinality of the domain of individuals (in the relative sense).

We now provide a simple characterization of tenability. A deeper account will later be provided in sect. 7.

Lemma 1 (Anti-inflation). Suppose that M is a standard §-free model and that ϕ is a §-free identity criterion. There is then an §-expansion M^+ of M that is a model of Φ iff:

  (i) $E_{\phi, M}$ is an equivalence relation; and

  (ii) $\text{card}(P_{\phi, M}) \leq \text{card}(M)$.

Proof. We first prove the direction from left to right. Suppose that the §-expansion M^+ of M is a model of Φ. By the Equivalence Lemma 3.3, E_ϕ is an equivalence relation. For each C in R^1, let $|C| = \{D: <C, D> \in E_\phi\}$ (so $P_\phi = \{|C|: C \in R^1\}$). Since M^+ is a model for Φ, $\S(C) = \S(D)$ iff $<C, D> \in E_\phi$; and so $\S$ induces a one-to-one map f from P_ϕ into M (with $f(|C|) = \S(C)$). But then card $(P_{\phi, M}) \leq \text{card}(M)$.

For the other direction, suppose that (i) and (ii) hold. Let $c = \text{card}(P_\phi)$; and pick a subset M' of M of cardinality c. There is then a one-to-one map f from P_ϕ onto M'. Define $\S$ by $\S(C) = f(|C|)$ for each $C \in R^1$. Then adding $\S$ as the abstractor to M yields an §-expansion M^+ that is a model of Φ.

The above proof does not rest on any special assumptions concerning the model M, although its formulation does require that ϕ be §-free. If we were to drop the requirement that M and ϕ be §-free, conditions (i) and (ii) would provide neither a necessary nor a sufficient condition for a variant of M (with possibly different §) to be a model of ϕ. They would not be jointly necessary since ϕ might be the trivial condition §$C = $§$D$. Nor would they be suffcient. For let ϕ be the formula $(\forall C, D(\S C = \S D) \,\&\, (\forall x Cx \leftrightarrow \forall x Dx)) \vee (\neg \forall C, D(\S C = \S D) \,\&\, C \equiv D)$. If we let card $(M) > 1$ and let § in M have a singleton range, conditions (i) and (ii) will be satisfied; and yet there is no model for Φ.

An analogous result can be proved for the existence of a model for Φ with exactly d individuals. In this case, condition (ii) should be replaced with:

(ii)$'$ $\mathbf{d} + \text{card}(P_{\phi, M}) = \text{card}(M)$.

If $\mathbf{m} = \text{card } (M)$ is infinite, then this condition is satisfied iff $\max\{\mathbf{d}, \text{card}(P_{\phi, M})\} = \mathbf{m}$.

Suppose that ϕ is an L-criterion. Then we say that the cardinal $\mathbf{c}$ is *non-inflationary* (*wrt* ϕ) if the conditions (i) and (ii)$'$ are satisfied when M is the standard model M_c. We now have the following immediate consequences of the lemma:

Corollary 2. For ϕ an L-criterion:
(i) ϕ is tenable on $\mathbf{c}$ iff $\mathbf{c}$ is non-inflationary;
(ii) ϕ is stable iff for some cardinal $\mathbf{c}$, every $\mathbf{d} \geq \mathbf{c}$ is non-inflationary.

An analogue to (i) holds for $\mathbf{d}$-tenability. Say that the cardinal $\mathbf{c}$ is *$\mathbf{d}$-static* if condition (i) of lemma 1 and the condition (ii)$'$ above are satisfied for the case in which M is the standard model M_c. Then ϕ will be $\mathbf{d}$-tenable on $\mathbf{c}$ iff $\mathbf{c}$ is $\mathbf{d}$-static. The matter might be put in terms of fixed points. For each cardinal $\mathbf{c}$, let the *partition cardinal* $p(\mathbf{c}) = \text{card}(P_{\phi, M})$ (for $M = M_c$ and $P_{\phi, M}$ defined) and let $p_d(\mathbf{c}) = \mathbf{d} + p(\mathbf{c})$. Then ϕ will be $\mathbf{d}$-tenable on $\mathbf{c}$ just in case $\mathbf{c}$ is a fixed point of the function p_d.

Given that ϕ is absolute, p_d will be a monotonic function—or, more exactly, $\mathbf{e} \leq \mathbf{f}$ and $p_d(\mathbf{f})$ defined will imply that $p_d(\mathbf{e})$ is defined and $p_d(\mathbf{e}) \leq p_d(\mathbf{f})$. From this it follows that, if p_d has a fixed point, then its least fixed point can be reached by iteration from below. Thus if we set $\mathbf{c}^0 = p_d(0)$, $\mathbf{c}^{\alpha+1} = p_d(\mathbf{c}^\alpha)$ and $\mathbf{c}^\lambda = \sup\{\mathbf{c}^\xi \colon \xi < \lambda\}$, then the least fixed point of p_d, if it exists, will be of the form $\mathbf{c}^\alpha$, for α the least ordinal for which $\mathbf{c}^\alpha = \mathbf{c}^{\alpha+1}$.

The definitions and results can be carried over to systems containing several abstraction operators. For example, when it is required that the model be separated, Lemma 1 takes the form:

Lemma 3 (Anti-inflation). Suppose that $\phi = \langle \phi_\S \colon \S \in \S \rangle$ is a system of §-free identity criteria and that M is a standard §-free model. There is then a separated §-expansion M^+ of M which is a model of the $\Phi_\S$ iff:
(i) each $E_{\phi, M}$, for $\phi = \phi_\S$, is an equivalence relation;
(ii) card $(\bigcup \{P_{\phi, M} \colon \phi = \phi_\S\}) \leq \text{card}(M)$.

Let us say that a system $\phi = \langle \phi_\S \colon \S \in \S \rangle$ of identity criteria is tenable on $\mathbf{c}$ if some separated expansion of the standard model M_c

is a model of all the $\phi_\S$ and that the system is *stable* if, for some cardinal **c**, it is tenable on every **d** ≥ **c**. We then have the corollary in the same form as before, but with non-inflation defined by means of the new conditions (i) and (ii) and with the system of criteria ϕ in place of the single criterion ϕ.

The tenability conditions can be expressed within second-order logic itself. This point will be important both for proving certain technical results and for setting up the general theory of abstraction in Part IV. In the first place, we may note that it follows from Lemma 3.3 that the abstraction principle Φ, corresponding to an identity criterion ϕ, will deductively imply the formula:

Eq_ϕ: $\forall C, D, E[\phi(C, C) \ \& \ (\phi(C, D) \rightarrow \phi(D, C)) \ \& \ (\phi(C, D) \ \& \ \phi(D, E) \rightarrow \phi(C, E)]$.

Thus we may show within second-order logic that it is necessary for an abstraction principle to hold that its identity criterion determine an equivalence relation on concepts.

We also wish to prove that it is necessary for an abstraction principle to hold that it not be 'inflationary'. Suppose again that the criterion of identity is ϕ. Then the condition that the principle not be inflationary can be expressed by means of the following third-order formula (where **R** is a variable ranging over second-order relations between concepts and objects):

(*) $\exists R[\forall C \exists! x(R(C, x) \ \& \ \forall C, D \forall x, y(R(C, x) \ \& \ R(D, y) \rightarrow (x = y \leftrightarrow \phi(C, D))]$.

It is clear that (*) follows within third-order logic from the abstraction principle Φ (i.e. $\forall C, D(\S C = \S D \leftrightarrow \phi(C, D))$); for we may take **R**(C,x) to be given by the condition $\S C = x$. Moreover, if ϕ is itself $\S$-free, then the formula (*) will also be $\S$-free.

Given an appropriate version of the axiom of choice for objects, (*) will be equivalent to:

(**) $\exists R[\forall C \exists x \ (R(C, x) \ \& \ \forall C, D \forall x(R(C, x) \ \& \ R(D, x) \rightarrow \phi(C, D))]$.

Moreover, given an appropriate version of the axiom of choice for concepts, (*) will, in its turn, be equivalent to the following second-order formula (where R is now a binary relation on objects):

$Noninfl_\phi$: $\exists R \forall C \exists D \exists y[\phi(C, D) \ \& \ \forall x(R(x, y) \leftrightarrow Dx)]$.

R in effect picks out a representative D from each equivalence class $|C|$.

Given the 'meaning' of Eq_ϕ and $Noninfl_\phi$ in a standard model M_c, it follows that:

Lemma 4. The L-criterion ϕ is tenable on c iff (Eq_ϕ & $Noninfl_\phi$) is true in the model M_c.

This result can be used to determine the least cardinal at which all stable criteria stabilize. Given an L-criterion ϕ that stabilizes, let c_ϕ be the cardinal at which it stabilizes; and let the *stability number* (*of* pure second-order logic) be the least upper bound of the c_ϕ. Recall that the *Hanf number* (*of* pure second-order logic) is the least cardinal such that any sentence with a standard model of that cardinality or greater has standard models of arbitrarily large cardinality. Let us extend the notion of stabilizing and say that an arbitrary *sentence* ψ of L^2 *stabilizes on* a cardinal c if c is the smallest cardinal with the property that ψ is true on all models of equal or greater cardinality. Then the Hanf number will be the least upper bound of the cardinals upon which a sentence of L^2 stabilizes. We now have:

Theorem 5. The Hanf number of second-order logic is identical to the stability number.

Proof. Let us use c for the Hanf number and c^* for the stability number. First suppose that $c < c^*$. Since c is less than the stability number, there is an L-criterion ϕ that stabilizes on some cardinal $d > c$. Let ψ be the sentence (Eq_ϕ & $Noninfl_\phi$). By Lemma 4, ψ stabilizes on d, contrary to assumption that c is the Hanf number.

Now suppose $c^* < c$. Since c^* is less than the Hanf number, it follows that there is a sentence ψ of L^2 that stabilizes on some cardinal $d > c^*$. Define the identity criterion ϕ as: ($\psi \lor (C \equiv D)$). Then it is readily shown that ϕ stabilizes on $d > c^*$, contrary to assumption that c^* is the stability number.

5. Generation

We deal here with the notion of a generative model, one in which the abstracts are generated in stepwise fashion from the individuals. The philosophical significance of the notion is discussed in sect. I.2 and I.3.

Let M be a standard §-model (with I, recall, as its subdomain of individuals). Define g on $\wp(M)$ by:

$$g(K) = \{\S(C): C \subseteq K\}.$$

Thus $g(K)$ consists of the abstracts that can be directly generated from the concepts that are defined over the objects of K. For each ordinal ξ, we define the *generation function* G_ξ by:

$$G_0 = I;$$
$$G_{\alpha+1} = I \cup g(G_\alpha);$$
$$G_\lambda = \bigcup\{G_\xi: \xi < \lambda\}.$$

We also set $g^+(X) = g(X) \cup X$. $G_{\alpha+1}$ can then be alternatively defined as $g^+(G_\alpha)$.

We prove in the usual way that there is a least ordinal α such that $G_{\alpha+1} = G_\alpha$ and call α the *critical* ordinal *for M*. We use G—without danger of confusion, I hope—for this G_α. Thus G is the set of objects that can be generated, either directly or indirectly, from the individuals of M.

The requirement that the objects be generable then amounts to the condition that, in the model M, $M = G$. We call call a model conforming to this condition *minimal*.

The condition that $M = G$ might be compared to the axiom of constructibility, $V = L$, or to the principle $V = \bigcup V_\alpha$, where the V_α's are the sets in the cumulative hierarchy. But the construction is in a way broader. For it relates to any kind of abstract, not just to sets; and the abstracts are generated by means of concepts, which may or may not be definable. One could, of course, carry out an analogous construction in a non-standard model M.

As an example of a minimal model, consider the standard model M in which the domain M is taken to consist of the natural numbers with $\aleph_0$ and $\S$ is the cardinal abstractor defined by: $\S(C) = \text{card}(C)$. Then M is generative. For: $G_0 = I = \phi; G_{n+1} = \{0, 1, \ldots, n\}; G_\omega = \{0, 1, \ldots\}$; and $G_{\omega+1} = \{0, 1, \ldots, \aleph_0\} = M$ and is therefore a fixed point. On the other hand, the standard model M in which M consists of all cardinals $\leq$ some inaccessible cardinal and $\S$ is the cardinal abstractor is not generative. For, again, I will be ϕ and the least fixed point G will be $\{0, 1, \ldots, \aleph_0\}$, which is a proper subset of M. In such a model, there will be no non-circular account of any of the cardinals $> \aleph_0$, for each of them must be explained by reference to a concept that holds of some of those very cardinals.

The definitions can be generalized. Say that a subset K of M is *downward-closed* if each element c of K is either an individual or is of the form $\S(C)$ for some subset C of K; say that K is *upward-closed* if, for any $C \subseteq K$, $\S(C) \in K$; and say that K is *closed simpliciter* if it is both downward- and upward-closed. The construction can then be relativized with an arbitrary downward-closed subset K of M in place of I. Each of the sets $G_{K,\alpha}$ obtained by means of the construction will then be downward-closed. Let G_K be the resulting least fixed point. Then G_K will also be upward-closed and hence closed. We say that the model M is *K-minimal* if $M = G_K$. Note that if a model to be is K-minimal then K must contain I.

Similar definitions can also be given with generalized §-models in place of the singular §-models. The set $g(K)$ must then contain $\S(C)$ for each subset C of K and abstractor $\S$ for which $\S \in \boldsymbol{\S}$.

We have the following alternative characterization of the sets G_K (proved in the usual way):

Lemma 1. Let M be a standard §-model and K a subset of M. Then G_K is the smallest upward-closed subset of M to contain K.

It follows that a model M is minimal if M is the smallest upward-closed subset of M to contain I. This alternative formulation of minimality can itself be expressed within the language of the abstraction theory. For let us use:

Closed (C) for $\forall x(Ix \to Cx)$ & $\forall D(D \subseteq C \to C\S(D))$; and
Min for $\neg \exists C(\exists x \neg Cx$ & Closed(C)).

We then see that:

Lemma 2. A standard model M is minimal iff $M \models$ Min.

Even in the absence of Min, we may define the smallest closed concept and show that it exists. For let $\psi = \psi(C, D)$ be the formula $\forall x(Dx \leftrightarrow \exists C'(C' \subseteq C \ \& \ x = \S C')$. Then, in any abstraction theory T^ϕ, we can prove the functionality and monotonicity conditions Func(ψ) and Mon(ψ) required for the application of the fixed-point theorem (Theorem 3.2); and so, by the fixed-point theorem itself, we can prove $\exists G \chi(G)$, where $\chi(G)$ is the formula $\exists C(\forall x(Cx \leftrightarrow Ix) \ \& \ \text{LFP}_{c,\psi}(G))$.

We may also express the generative account of closure within the system by using the following sequence of definitions:

Well-foundedness: WF(R) for $\forall C(\exists yCy \rightarrow \exists y(Cy \ \& \ -\exists x(Cx \ \& \ xRy))$;

Generating relation: Gen(R) for [WF(R) $\&$ $\forall y(y \in \text{Fld}(R)$ $\&$ $\exists C(y = \S C) \rightarrow \exists C(y = \S C \ \& \ \forall x(Cx \leftrightarrow xRy)))$];

Generable objects: Gen(x) for $\exists R(\text{Gen}(R) \ \& \ x \in \text{Fld}(R))$.

A generating relation corresponds to a construction of abstracts, with each abstract in the field of the relation being generated from a concept that holds of previously generated objects. It may be shown that the generable objects constitute the smallest closed domain (containing all individuals). But this requires the use of an appropriate form of the axiom of choice.

Another requirement that is natural from the generative standpoint is that each identity criterion ϕ be absolute in the sense of sect. 3. If it were not, then we would not properly know which abstract was being introduced by a definition of the form §C; for which abstract was denoted would depend not only upon the extension of the concept C but also upon what the underlying domain would ultimately turn out to be. Suppose, for example, that § were interpreted as the bicardinality operator, i.e. as an operator that took each concept into the pair consisting of the cardinality of its extension and the cardinality of its counter-extension. Then it would not be known which 'bicardinal' resulted from the application of § to an empty concept until it had been determined how many objects there would be in the domain.

The requirement of absoluteness is not as restrictive as it would appear. For suppose we are dealing with an abstract, like bicardinality, whose identity depends upon the underlying domain. We can then trade it in for an abstract on two concepts, where one concept corresponds to the original concept and the other to the domain. The new abstract of C and D will then correspond to the old abstract of C in the domain D. But of course this approach, once it is generalized, requires that we consider abstraction on any finite number of concepts.

In the case of absolute identity criteria, the existence of a minimal submodel is always guaranteed:

Theorem 3. For ϕ an absolute L-criterion, the minimal submodel *N* of the standard model *M* will be a model of the theory T^ϕ as long as *M* itself is.

Proof. Let *C* and *D* be any two subsets of *N*. Then:

$\S_N(C) = \S_N(D)$ iff $\S_M(C) = \S_M(D)$ (since N is a submodel of M)
iff $M \models \phi[C, D]$ (given that M is a model for T^ϕ)
iff $N \models \phi[C, D]$ (given that ϕ is absolute).

A syntactic form of this result may also be established. For where $\phi = \phi(C, D)$ is an identity criterion, let Absolute(ϕ) be the formula $\forall C\{\forall D, E(D, E \subseteq C \to (\phi^C(D, E) \leftrightarrow \phi(D, E)))\}$. Thus Absolute($\phi$) states that ϕ is absolute (though not absolutely absolute, only absolute relative to the underlying domain!). We may now prove within T^ϕ the formula: Absolute(ϕ) $\to \forall C(\text{Closed}(C) \to \Phi^C)$. Let Min(C) be the formula $\exists C(\text{Closed}(C) \ \& \ \forall D(\text{Closed}(D) \to C \subseteq D))$. Then we may also prove $\exists C \text{Min}(C)$. Hence given a proof of Absolute(ϕ), we may show $\exists C(\text{Min}(C) \ \& \ \Phi^C)$. The formula that defines C may be regarded as an 'inner model' for the theory T^ϕ.

The theorem can be extended to generalized theories T^ϕ in which each $\phi_\S$ is an absolute L-criterion. A version of the result can also be proved for generalized theories T^ϕ defined on a well-ordered sequence $<\S_\xi: \xi < \alpha>$ of abstraction operators (i.e. theories in which $\phi_\S$, for $\S = \S_\xi$, only contains the operators $\S_\nu$ for $\nu < \xi$). In this case, we require that M and ϕ conform to the condition that, for any $K \subseteq M$ and any $C, D \subseteq K, M \models \phi_\xi[C, D]$ iff $M/K \models \phi_\xi[C, D]$ as long as M/K is a model of each Φ_ν for $\nu < \xi$.

The condition that ϕ be absolute is required for the theorem to hold. Suppose, for example, that ϕ is the criterion of biequinumerosity and that M is the model for T^ϕ in which $M = \{ <\mathbf{c}, \mathbf{d}>: \mathbf{c} + \mathbf{d} = \aleph_0\}$ and $\S(C) = <\text{card}(C), \text{card } (M - C)>$ for each $C \subseteq M$. Then the minimal submodel N will be the restriction of M to $\{ <\mathbf{c}, \aleph_0>: \mathbf{c} \leq \aleph_0\}$, which is not a model for T^ϕ. Indeed, in this case it can be shown that the theory T^ϕ has no standard minimal model.

6. Categoricity

We prove a basic categoricity result and then outline various ways in which it might be extended. The relevance of categoricity to the question of definition has been discussed in sects. I.2 and I.3. We assume throughout this section that the underlying languages contain no non-logical symbols apart from the abstraction operators.

Our basic categoricity result will depend upon two lemmas concerning extensions and chains. We say that two standard $\S$-models M

and N are *internally similar* if, for any subsets C and D of M, subsets C' and D' of N, and one-to-one map f for which $f[C] = C'$ and $f[D] = D'$, $\S_M(C) = \S_M(D)$ iff $\S_N(C') = \S_N(D')$. Thus whether two subsets are identified by $\S$ in either model depends upon their possessing the same internal structural relationships.

Lemma 1 (Extension). Let M and N be two internally similar standard models; and let K and L be two downward-closed subsets of the respective domains M and N. Suppose that f is an isomorphism from M/K onto N/L. Then f can then be extended to an isomorphism f^+ from $M/g^+(K)$ onto $N/g^+(L)$.

Proof. It should be noted that the various restrictions M/K, $N/g^+(L)$, etc., will in general be partial models. We extend f to a function f^+ on $g^+(K)$ by letting $f^+(\S_M(C))$ be $\S_N(f[C])$ for each subset C of K for which $\S_M(C)$ is not in K. We first need to show that f^+ is well defined, which is to say that $\S_N(f[C])$ and $\S_N(f[D])$ are the same whenever $\S_M(C)$ and $\S_M(D)$ are the same (and not in K). But this follows from the fact that f is one-to-one and the models are internally similar.

We now show that f^+ is one-to-one. To this end, we show that $f^+(c)$ and $f^+(d)$ are distinct for distinct $c, d \in g^+(K)$. We distinguish three cases:

(a) $c, d \in K$. Then $f^+(c) = f(c)$ and $f^+(d) = f(d)$ and $f(c)$ and $f(d)$ are distinct by f an isomorphism.

(b) $c, d \in g(K)$. Then c is of the form $\S_M(C)$ and d is of the form $\S_M(D)$ for some subsets C and D of K. By the definition of f^+ and the fact that f is an isomorphism, $f^+(\S_M(C)) = \S_N(f[C])$ and $f^+(\S_M(D)) = \S_N(f[D])$; and given that $c = \S_M(C)$ and $d = \S_M(D)$ are distinct, it follows from internal similarity that $\S_N(f[C])$ and $\S_N(f[D])$ are distinct.

(c) $c \in K - g(K)$ and $d \in g(K) - K$ (and similarly when $d \in K - g(K)$ and $c \in g(K) - K$). Then $f^+(d)$ is of the form $\S_N(D')$ for some subset D' of L, where D' is itself of the form $f(D)$ for some subset D of K. Suppose, for reductio, that $f^+(c) = \S_N(D')$. Then since $f^+(c) = f(c)$, it follows from f an isomorphism that $c = \S_M(D)$; and hence $c \in g(K)$ after all.

We next show that f^+ is onto. Suppose $c' \in g^+(L)$. If $c' \in L$, then c' is in the range of f and hence of f^+. So suppose $c' \in g(L) - L$. Then c' is of the form $\S_N(C')$ for some subset C' of L. But then $f^+(\S_M(f^{-1}[C'])) = c'$.

Finally we show that, for any subset C of $g^+(K)$, $\S_M(C)$ is defined iff $\S_N(f^+[C])$ is defined and that, in case they are both defined, $f^+(\S_M(C)) = \S_N(f^+[C])$. Let us suppose that $\S_M(C)$ is defined and identical to c (the other direction is similar). Then:

(*) $\S_M(C) = \S_M(C')$ for some subset C' of K.

For either $c \in g^+(K) - K$, in which case (*) holds by the definition of g^+, or $c \in K$, in which case (*) holds by the fact that K is downward-closed. From (*), $f^+(\S_M(C)) = f^+(\S_M(C')) = \S_N(f[C'])$ $= \S_N(f^+[C'])$; and, by internal similarity, $\S_N(f^+[C]) = \S_N(f^+[C'])$.

Note, for the subsequent application of this lemma, that $g^+(K)$ will be downward-closed whenever K is.

Lemma 2 (Chains). For λ a limit ordinal, let $<M/K_\xi : \xi < \lambda>$ and $<N/L_\xi : \xi < \lambda>$ be two increasing chains of partial $\S$-models and $<f_\xi : \xi < \lambda>$ an increasing chain of functions (i.e. $K_\xi \subseteq K_{\xi'}$, $L_\xi \subseteq L_{\xi'}$, and $f_\xi \subseteq f_{\xi'}$ for $\xi < \xi' < \lambda$). Suppose that f_ξ is an isomorphism from M/K_ξ onto N/L_ξ for each $\xi < \lambda$. Then $f = \bigcup f_\xi$ is an isomorphism from $M/\bigcup K_\xi$ onto $N/\bigcup L_\xi$.
Proof. A straightforward verification.

Also note, for purposes of subsequent application, that the limit $<M/K_\xi : \xi < \lambda>$ will be downward-closed whenever each of the models M/K_ξ is downward-closed.

Theorem 3 (Categorical Extension). Suppose that M is a standard K-minimal $\S$-model for K a downward-closed subset of M and that N is an internally similar standard L-minimal $\S$-model for L a downward-closed subset of N. Suppose that f is an isomorphism from M/K onto N/L. Then there exists a unique extension f^+ of f to an isomorphism from M onto N.
Proof. Let G and H be the respective generation functions for M and N (relativized to K and L respectively). For each ordinal ξ, we define an isomorphism f_ξ from M/G onto N/H as follows:

(i) For $\xi = 0$, we let f_ξ be f. f_ξ is then by supposition, an isomorphism from M/G_0 onto N/H_0.

(ii) For $\xi = \alpha + 1$, we let f_ξ be an extension of f_α which is an isomorphism from M/G_ξ onto N/H_ξ. Since (by inductive hypothesis) we may assume that f_α is an isomorphism from

M/G_α onto N/H_α and, given that G_α is downward-closed, such an isomorphism will exist by the extension lemma.

(iii) For $\xi = \lambda$, we let f_λ be the union of the f_ξ for $\xi < \lambda$. By the chain lemma, f_λ will again be an isomorphism from M/G_λ onto N/H_λ.

Since both M and N are minimal, we may choose an ordinal α for which $G_\alpha = M$ and $H_\alpha = N$. But then $f^+ = f_\alpha$ is an isomorphism from M onto N.

To show that the extension f^+ is unique, suppose that there is another extension f^* of f that is an isomorphism from M onto N. Choose the least α for which there exists an element c in G_α upon which f^+ and f^* disagree. Clearly $\alpha > 0$. But then c is of the form $\S_M(C)$ for C a subset of G_β for some $\beta < \alpha$. Now $f^+(\S_M(C)) = \S_N(f^+[C])$ and $f^*(\S_M(C)) = \S_N(f^*[C])$. Since $\beta < \alpha$, $f^+[C] = f^*[C]$; and so $f^+(c) = f^*(c)$ after all.

As a special case of the theorem, we obtain:

Corollary 1. Suppose that M is a standard minimal §-model with invariant $\equiv_\S$. There are then no proper automorphisms on M that are fixed on I_M.

Proof. Since $\equiv_\S$ is I-invariant, M is internally similar with itself. Suppose that f is the identity on I_M. Then, by the theorem, there exists a unique extension f^+ of f that is an automorphism on M.

Thus, once the identity of the individuals is taken to be given, each abstract will have its own structural 'position'.

As a further special case of the theorem, we have:

Corollary 5. Suppose that M and N are internally similar standard minimal §-models.

Then they are isomorphic as long as card $(I_M) = \text{card}(I_N)$.

Proof. Since card$(I_M) = \text{card}(I_N)$, there is a one-to-one map f from I_M onto I_N. Clearly, I_M and I_N are both downward-closed and f is an isomorphism from M/I_N. But then M and N are isomorphic by the theorem.

We do not have this result without the requirement that both M and N be minimal, even when we fix the cardinality of the subdomains of individuals and of abstracts. For let M be a standard model with domain $\{a^0, a^1, \ldots\}$, $\S(\{a^n\}) = a^{n+1}$ for $n \geq 0$, and $\S(C) = a^0$ for C non-singleton; and let N be a standard model with domain

$\{a^0, a^1, \dots\} \cup \{\dots, b^{-1}, b^0, b^1, \dots\}$, $\S(\{a^n\}) = a^{n+1}$ for $n \geq 0$, $\S$ (C) $= a^0$ for C non-singleton, and $\S(b^n) = b^{n+1}$ for any integer n. Then M and N are internally similar; and both have zero individuals and denumerably many abstracts. But they are not isomorphic. Indeed, M is minimal; whereas N is not, since it contains a 'backwards' regress with respect to $\S$.

There is, however, a circumstance in which cardinality alone is sufficient to secure isomorphism. Let us say that a §-model M is *numerical* if $\S(C) = \S(D)$ whenever C and D are two subsets of M of the same cardinality.

Theorem 6. Suppose that M and N are internally similar standard numerical §-models of the same cardinality. Then they are isomorphic as long as $\operatorname{card}(I_M) = \operatorname{card}(I_N)$.

Proof. In this case we may define the isomorphism f directly by letting f/I_M be a one-to-one map onto I_N and by letting $f(\S_M(C)) = \S_N(C')$ for C' a subset of N of the same cardinality as C.

The proof of Theorem 6 actually requires only that the models be binumerical in the sense of identifying concepts with the same bicardinality and that they be exactly similar in the sense that, for any one-to-one map f from M onto N, $\S_M(C) = \S_M(D)$ iff $\S_N(f[C]) = \S_M(f[D])$.

Corollary 5 has an immediate consequence for theories:

Corollary 7 (Categoricity). Suppose that ϕ is an absolute L-criterion. Then any two standard models for $T^\phi +$ Min with the same number of individuals are isomorphic.

Proof. Given that ϕ is absolute, it follows that any two standard models M and N of T^ϕ will be internally similar. It therefore follows from the corollary that any two such models will be isomorphic as long as the cardinalities of I_M and I_N are the same.

Thus the combination of absoluteness and minimality is able to secure categoricity relative to the cardinality of the individuals. The cardinality of the minimal model with $\mathbf{d}$ individuals will be the least fixed point $\mathbf{c}$ of the function $p_\mathbf{d}$ defined in sect. 4. For ϕ will have a model with exactly $\mathbf{c}$ objects and $\mathbf{d}$ individuals, and so will have a generated submodel, which has at most $\mathbf{c}$ objects and exactly $\mathbf{d}$ individuals, and hence which has exactly $\mathbf{c}$ objects given that $\mathbf{c}$ is the least fixed point.

In the case of numerical identity criteria (those which determine numerical §-models), any formula that fixes the cardinality of the abstracts will do in place of Min. Suppose, for example, that we define *Natural Number* in the abstraction theory with Hume's Law. Then any axiom to the effect that there is a one-to-one correspondence between the universe and the natural numbers will be sufficient to secure (relative) categoricity, as will any axiom to the effect that there is a one-to-one correspondence between the universe and all sets of natural numbers. Boolos (1987: 16) and Dummett (1991 a: 227) have both commented on the failure of Hume's Law to settle questions of cardinality. But this, we see, is the only respect in which it fails to achieve completeness or, indeed, categoricity.

The generation of the minimal model, in these cases, precedes in an especially simply way. Given $v < \xi < \alpha$, where α is the critical ordinal, $card(G_v) < card(G_\xi)$. For clearly the result will hold if it holds for $\xi = v + 1$. So suppose, for reductio, that $card(G_v) = card(G_{v+1})$. Then $v + 1$ must be the critical ordinal. For take an abstract $\S(C)$ in G_{v+2}, with $C \subseteq G_{v+1}$. Choose a subset C' of G_v of the same bicardinality as C. Then $\S(C') = \S(C)$; and hence no new abstracts are introduced at stage $v + 1$.

Let us now consider some extensions of the basic result to generalized theories of abstraction. Different results are obtained according as to whether or not we insist on the identity criteria being logical or on the models being strictly separated. In each case, the abstractors are taken to be denoted by an abstraction operator; and it is only in sect. IV.2 that we consider the problem of categoricity for a theory in which we are allowed to quantify over the methods of abstraction themselves.

The first extension is to generalized models in which the abstractors are taken to be logical but separated. Let M and N be two generalized §-models. We say that they are *internally similar* if the §-models $M_\S$ and $N_\S$ are internally similar for each § $\in$ §.

Theorem 8. Suppose that M is a strictly separated and standard K-minimal §-model, with K a downward-closed subset of M, and that N is a strictly separated and standard L-minimal §-model, with L a downward-closed subset of N. Suppose that M and N are internally similar and that f is an isomorphism from M/K onto N/L. Then there exists a unique extension f^+ of f to an isomorphism from M onto N. In particular, two internally similar models M and N will be isomorphic as long as $card(I_M) = card(I_N)$.

Proof. Form an ordered set $<\S_\zeta: \zeta < \alpha_0>$ of the operators $\S \in \mathbf{\S}$. For each ordinal ξ, define an isomorphism f_ξ from $M_\xi = M/K_\xi$ onto $N_\xi = N/L_\xi$ as follows:

(i) $\xi = 0$. Let $K_0 = K$, $L_0 = L$, and $f_\xi = f$. Then by supposition, f is an isomorphism from M_ξ onto N_ξ.

(ii) ξ of the form $\alpha_0\beta + \zeta'$ for $\zeta < \alpha_0$ (where ζ' is the successor of ζ). Let ν be the immediate predecessor $\alpha_0\beta + \zeta$ of ξ; and temporarily set $\S = \S_\zeta$. Recall that $M_\S$ and $N_\S$ are the respective reducts of M and N wrt $\S$. By Theorem 3 and IH, there is an extension $(f_\nu)^+$ of f_ν which is an isomorphism from a closed submodel $M_\S/K^*$ of $M_\S$ onto a closed submodel $N_\S/L^*$ of $N_\S$. We then set $K_\xi = K^*, L_\xi = L^*$, and $f_\xi = (f_\nu)^+$.

(iii) ξ of the form $\alpha_0\beta + \lambda$ for λ a limit ordinal $< \alpha_0$. We let K_ξ, L_ξ, and f_λ be the respective unions of the K_ν, L_ν, and f_ν for all $\nu < \xi$.

Using this construction, the results may then be proved in much the same way as before. Note that the models M and N are required to be strictly separated in order to guarantee that no new $\S'$-abstracts, for $\S' \neq \S$, will be introduced at step (ii).

We turn next to the case in which the operators are allowed to be defined in terms of one another. Let $<\S_\zeta: \zeta < \alpha>$ be a well-ordered sequence of the operators $\S \in \mathbf{\S}$ (which we regard as respecting their order of definition). Given such a sequence, an ordinal $\zeta < \alpha$, and a $\mathbf{\S}$-model M, we let $M_{<\zeta}$ be the reduct of M to the operators $\S_{\zeta'}$, with $\zeta' < \zeta$, and we say that the subset K of M is *closed$_{<\zeta}$* if it is closed wrt each operator $\S_{\zeta'}$ for $\zeta' < \zeta$. We say that the $\mathbf{\S}$-models M and N are *internally similar (relative to $<\S_\zeta: \zeta < \alpha>$)* if, for each $\zeta < \alpha$, for any isomorphism f from $M_{<\zeta}/K$ onto $N_{<\zeta}/L$, with K and L both closed$_{<\zeta}$, and for any subsets C, $D \subseteq K$ and C', $D' \subseteq L$ for which $f[C] = C'$ and $f[D] = D'$, $\S_{\zeta, M}(C) = \S_{\zeta, M}(D)$ iff $\S_{\zeta, N}(C') = \S_{\zeta, N}(D')$. Thus the map f is now required to be a full isomorphism on the preceding abstractors.

Theorem 9. Suppose that M is a strictly separated and standard K-minimal $\mathbf{\S}$-model, with K a downward-closed subset of M, and that N is a strictly separated and standard L-mimimal $\mathbf{\S}$-model, with L a downward-closed subset of N. Suppose that M and N are internally similar relative to the well-ordered sequence $<\S_\zeta: \zeta < \alpha>$ of the operators $\S \in \mathbf{\S}$ and that f is an isomorphism from M/K onto N/L.

There then exists a unique extension f^+ of f to an isomorphism from M onto N. In particular, the two models M and N will be isomorphic as long as $\text{card}(I_M) = \text{card}(I_N)$.

Proof. By transfinite induction on α. For $\alpha = 0$, the result is trivial. So suppose α is a successor ordinal $\beta + 1$. We then prove the extension lemma in the form:

(*) Let M and N be two strictly separated standard models that are internally similar relative to the sequence $< \S_\zeta: \zeta < \alpha >$; and let K and L be subsets of the respective domains M and N, closed wrt each $\S_\xi$, for $\xi < \beta$, and downward-closed wrt $\S_\beta$. Suppose that f is an isomorphism from M/K onto N/L. Then f can be extended to an isomorphism f^+ from $M/g^+(K)$ onto $N/g^+(L)$ (with g^+ defined in terms of $\S_\beta$).

The proof follows that of Lemma 1. Note that in the former proof an appeal to internal similarity is made at two places: the first is at (b) under the verification of one-to-oneness; and the second is at the end under the verification of the condition of isomorphism. The first gives rise to no difficulty; but the second does, since the application of the internal similarity condition in its relativized form now requires that we have a suitable underlying isomorphism. To overcome this difficulty, we appeal to the inductive hypothesis. For the f^+, as defined under (*), can be extended to an isomorphism from $M_{<\beta}/K^+$ onto $N_{<\beta}/L^+$, where K^+ and L^+ respectively contain $g^+(K)$ and $g^+(L)$ and are closed$_{<\beta}$.

Having established the extension lemma, we show, along the lines of the proof of Theorem 3, that there is an extension f^+ of f (in the statement of the present theorem) that is an isomorphism from M_β/K^+ onto N_β/L^+, where K^+ and L^+ respectively contain K and L and, by strict separation, contain no new $\S_\zeta$-abstracts, for $\zeta < \beta$. The inductive hypothesis may now be applied, with K^+ and L^+ in place of K and L. This may, in its turn, be followed by an extension to take care of the operator $\S_\beta$—and so on, until a fixed point is obtained.

Finally suppose that α is a limit ordinal λ. For each $\xi < \lambda$, let f_ξ be the unique extension of f which is an isomorphism from $M_{<\xi}/K_\xi$ onto $N_{<\xi}/L_\xi$ with K_ξ and L_ξ closed$_{<\xi}$ supersets of K and L. Then it is clear that the f_ξ's, for $\xi < \lambda$, form an increasing chain; and so $f^+ = \bigcup f_\xi$ will be an isomorphism from $M = M/\bigcup K_\xi$ onto $N = N/\bigcup L_\xi$.

We can apply this result to particular generalized theories by extending the notion of an absolute axiom. Instead of requiring that the 'meaning' of the identity criterion $\phi(C,D)$ remain the same under the relativization to any concept that subsumes C and D, we now require that the meaning remain the same under the relativization to any concept that subsumes C and D and is closed under the preceding abstraction operations.

We turn finally to the question of categoricity for separated, though not strictly separated, §-models. In this case, the analogue of Theorem 8 does not hold. For consider a § with two members $\S_1$ and $\S_2$. Let M be a standard §-model with $M = \{c\}$ and $\S_{1,M}(C) = \S_{2,M}(C) = c$ for all subsets C of M; and let N be a standard model with $N = \{c, d, e\}, \S_{1,N}(C) = c$ for all subsets C of N, $\S_{2,N}(C) = d$ for empty and singleton subsets C of N, and $\S_{2,N}(C) = e$ for doubleton subsets C of N. Then M and N are separated minimal models without individuals, satisfy the internal similarity condition, and yet are not isomorphic.

The discrepancy in the generation of M and N emerges at the first stage. For whereas $\S_{1,M}$ and $\S_{2,M}$ yield the same abstract c in application to the empty concept, $\S_{1,N}$ and $\S_{2,N}$ yield the different abstracts c and d. There is no violation of separation in the case of N since the abstracts c and d can be distinguished in terms of concepts defined with the help of those very abstracts.

It is reasonable to think that the model N should not be allowed under a proper conception of the generative method; for at the first stage, we have no basis in terms of the objects already generated, for distinguishing between the abstracts c and d. Accordingly, let us redefine the notions of generation and minimality.

Let K be a subset of the domain M of a standard §-model M. For $\S, \S' \in \S$, we say that the abstract $\S'(C)$ is *indistinguishable from* the abstract $\S(C)$ *in* K if $C \subseteq K$ and $\{D \subseteq K : \S'(D) = \S'(C)\} = \{D \subseteq K : \S(D) = \S(C)\}$. (Note that the indistinguishability of the two abstracts depends upon their respective representation as $\S$- and $\S'$-abstracts). We now say that the subset K of M is *quasi-upward-closed* if it contains I and if for any subset C of K and abstractor $\S$, with $\S \in \S$, there is an abstractor $\S'$, with $\S' \in \S$, such that $\S(C)$ is indistinguishable from $\S'(C)$ in K and $\S'(C) \in K$. Thus we no longer require that $\S(C)$ itself be in K but only that some indistinguishable abstract be in K. We do not insist on the stronger condition of upward-closure since there is no basis from within K for distinguishing between indistinguishable abstracts.

The generation of a minimal upward-closed subset of a domain M is no longer straightforward since, at a given stage in the generation of the subset, we may have a choice as to which of two currently indistinguishable abstracts should be introduced. Accordingly, the previous generation function g should be replaced by a system of alternative generation functions. Accordingly, let us say that g is *a generation function for* the standard §-model M if, for every subset K of M, $K^+ = K \bigcup g(K)$ is a set with the properties:

(i) for each $C \subseteq K$ and $\S \in \S$, there is a $\S' \in \S$ such that $\S(C)$ is indistinguishable from $\S'(C)$ in K and $\S'(C) \in g(K)$;

(ii) any object $c \in g(K)$ is of the form $\S'(C)$ for $C \subseteq K$ and is indistinguishable in K from any other abstract $\S(C)$ in $K \cup g(K)$. Thus the $g(K)$ may be obtained by first forming the indistinguishability classes of the $\S(C)$, for $\S \in \S$ and $C \subseteq K$, and then picking a representative from those classes which do not already have a representative in K.

The generation of the upward-closed subset may now be relativized to a generation function g in the usual manner. Thus for any ordinal ξ:

$$G_0 = I;$$
$$G_{\alpha+1} = I \cup g(G_\alpha);$$
$$G_\lambda = \bigcup \{G_\xi : \ \xi < \lambda\}.$$

By considerations of cardinality, there will be a least ordinal α for which $G_{\alpha+1} = G_\alpha$. We again call α the *critical* ordinal *for* M and again use G for G_α. It is evident that G is quasi-upward-closed and that it is separated in the sense that any two abstracts $\S(C)$ and $\S'(C)$ in G are distinguishable in G.

Let us say that the §-model M is *strictly minimal* if it is minimal and if M properly contains no quasi-upward-closed subset. As is evident from the example N above, there is no guarantee that even an internally invariant §-model will contain a strictly minimal submodel. However, should such a submodel exist, it may be generated by means of any one of the generation functions g. In particular, a strictly minimal model may itself be generated by means of such a function and hence the objects of such a model may be generated in such a manner that the abstracts introduced at any given stage are always distinguishable.

Using the condition of strict minimality in place of minimality, we may extend our previous categoricity results to separated §-models. The proofs are much the same as before, but instead of considering an arbitrary generation of the objects of the model we consider a generation that is in accord with one of the generation functions g.

It is natural to think of each generated abstract as being represented by a term. Given such a representation, one may then provide a truth-definition for the sentences of an abstraction theory in quasi-nominalistic terms, without reference to an underlying domain of objects. Such a truth-definition is very much in accord with the kind of truth-conditions that the proponent of the context principle might be thought to have in mind. However, he will be unable to offer the truth-definition that is stated here since it requires mathematical resources that go well beyond those at his disposal.

We suppose given a set of *individual constants* $\{a_0, a_1, \ldots\}$ of arbitrary cardinality, which we regard as denoting the individuals— one constant for each individual. We define the *object* and the *relation* terms by the following rules:

(i) each individual constant is an object term;
(ii) each set of n-tples of object terms is a relation term (in particular, a set of object terms is said to be a *concept* term);
(iii) If C is a concept term, then §C is an object term.
Note that the object and the concept terms will each form a proper class.

We think of the concept term $\{t_1, t_2, \ldots\}$ as denoting the concept whose extension consists of the denotations of $t_1, t_2, \ldots$ (and similarly for the relation terms). We have represented a concept term by a set, but we could equally well have represented it by something more linguistic in character, such as the 'expression' $\lambda x(x = t_1 \lor x = t_2 \lor \ldots)$. The infinitary character of the concept terms reflects the fact that we think of the concepts as being given platonistically.

We think of the object term §C as denoting the abstract on the concept denoted by C. There will, of course, be no assurance that different object terms §C denote different abstracts and hence no assurance that the different concept terms denote different concepts. To each object term t, we associate a *rank* rk(t) according to the rules:

(i) $rk(a) = 0$ for a an individual constant;

(ii) $rk(\S C) = \sup\{rk(s) + 1: s \in C\}$.

Similarly, the rank of a set T of object terms is taken to be $\sup\{rk(t) + 1: t \in T\}$.

The new object and relation terms may be used to expand the language $L^{\S}$ of abstraction theory. To this end, we define the notions of *objectual term, relational term,* and *formula*:

(i) any objectual variable or object term is an objectual term;

(ii) if C is a conceptual term, then $\S C$ is an objectual term;

(iii) any n-ary relational variable or relation term is a relational term;

(iv) $s = t$ is a formula when s and t are objectual terms;

(v) $Pt_1 \ldots t_n$ is a formula if P is an n-ary relational term and $t_1, \ldots, t_n$ are objectual terms;

(vi) $\neg\phi$ and $(\phi \vee \psi)$ are formulas if ϕ and ψ are formulas;

(vii) $\forall x\phi$ is a formula when x is an objectual variable and ϕ is a formula; and

(viii) $\forall P\phi$ is a formula when P is a relational variable and ϕ is a formula.

A set of object terms T is said to be a (*term*) *domain* if $s \in T$ whenever $\S C \in T$ and $s \in C$. Thus a domain must contain the object terms that figure in the concept terms from which the object terms in the domain are formed. Given any set C of object terms, there will be a smallest domain T_c to contain C. With each closed sentence ϕ of the expanded language, we may associate the set $L(\phi)$ of object terms that occur in ϕ and the corresponding domain $T_\phi = T_{L(\phi)}$.

We define truth for the (closed) sentences of the expanded language. In fact, for the purposes of the induction, we need to define truth relative to a term domain T. It is assumed that this notion of relative truth has application only to a sentence ϕ and a domain T when $T_\phi \subseteq T$.

We suppose that we have been supplied with a specific identity criterion $\phi(C, D)$ for the operator $\S$. The various definitions and results are then relative to that specific choice of $\phi(C, D)$. The relative truth-definition is as follows:

(i) $a = b$ is true in T iff $a = b$, where a and b are individual constants;

(ii) $a = \S C$ for a an individual constant is never true in any domain T;

(iii) $\S C = \S D$ is true in T iff $\phi(C, D)$ is true in $T_{C \cup D}$;

(iv) $Pt_1 \ldots t_n$ is true in T iff for some $s_1, \ldots, s_n$, $t_1 = s_1, \ldots, t_n = s_n$ are true in T and $<s_1, \ldots, s_n> \in P$, for P an n-ary relation term;

(v) $\neg\phi$ is true in T iff ϕ is not true in T;

(vi) $(\phi \vee \psi)$ is true in T iff ϕ is true in T or ψ is true in T;

(vii) $\forall x \phi(x)$ is true in T iff $\phi(t)$ is true for every object term t in T;

(viii) $\forall P^n \phi(P^n)$ is true in T iff $\phi(Q)$ is true for every relation term $Q \subseteq T^n$.

The clauses are analogous to the clauses of a standard substitutional semantics, with the sole exception of clause (iii). Here, instead of adopting:

(iii)$'$ $\S C = \S D$ is true in T iff $\phi(C, D)$ is true in T,

we restrict the domain on the right-hand side. In effect, we are assuming that $\S C = \S D$ is true in the domain T iff it is true in the subdomain $T_{C \cup D}$. We shall say that an identity $s = t$ is *true (simpliciter)* if it is true in the domain $T_{\{s, t\}}$.

Without this modification to (iii)$'$, the truth-definition would not be well founded. But with the modification, it is. For let the *complexity of* a pair $<T, \phi>$, for ϕ a formula and T a domain containing T_ϕ, be the pair $<\alpha, n>$, where α is the rank of T and n is the number of occurrences of logical symbols in ϕ. Let us adopt a lexicographic ordering on measures of complexity; and let us suppose that clause (iv) is rewritten so that applications of clauses (i)–(iii) are already built into the evaluation of the right-hand side. Then it may be shown that the application of each clause to a formula results in a drop in complexity or a direct assignment of truth-value.

Let $Eq_=$ be the formula $\forall x\ (x = x)$ & $\forall xy\ (x = y \rightarrow y = x)$ & $\forall xyz\ (x = y\ \&\ y = z \rightarrow x = z)$ (contrast with the formula Eq_ϕ, which states that the criterion $\phi = \phi(C, D)$ is an equivalence). Given the clauses for $=$ in the truth-definition, there is no guarantee that $Eq_=$ will be true in any given term domain T, since that will depend upon the choice of the criterion $\phi(C, D)$. But let us assume that it is, and define $\approx_T$ by: $s \approx_T t$ iff s and t are terms in T and $s = t$ is true. Then $\approx_T$ will be an equivalence relation. So with each term $t \in T$, we may associate an equivalence class $|t| = \{u: u \approx_T t\}$. Let

$M_T = \{|t|: t \in T\}$; for each $Q \subseteq T^n$, let $|Q| = \{<|t_1|, \ldots, |t_n|> : <t_1, \ldots, t_n> \in Q\}$; and, where $T = T_c$, let M_c designate the standard second-order model with domain M_T. We use the subscripted notation $|t|_c$ and $|Q|_c$, where necessary, to indicate that the equivalence classes are with respect to the relation $\approx_T$.

Lemma 10. Suppose that $Eq_=$ is true in a term domain T. Let $\psi = \psi(x_1, \ldots, x_m, P_1, \ldots, P_n)$ be a formula of L^2 with free variables as displayed; and let $t_1, \ldots, t_m$ be object terms of T and $Q_1, \ldots, Q_n$ be relation terms of the same arity, respectively, as $P_1, \ldots, P_n$. Then $\psi(t_1, \ldots, t_m, Q_1, \ldots, Q_n)$ is true in T iff $\psi [|t_1|, \ldots, |t_m|, |Q_1|, \ldots, |Q_n|]$ is true in M_T.

Proof. By a straightforward induction on the truth-definition, treating the truth-values of the identity sentences as given.

It should be noted, from the proof of this result, that nothing turns upon the rules of evaluation for the identity sentences.

Lemma 11. Suppose that $Eq_=$ is true in the domain T. Let S be a subdomain of T; let N be the restriction of M_T to $\{|s|_T: s \in S\}$; and, for each $s \in S$, let $f(|s|_T) = |s|_T \cap S$. Then f is an isomorphism from N onto M_s.

Proof. A straightforward verification.

When can we expect $Eq_=$ to be true in a term domain? Call an L-criterion $\phi = \phi(C, D)$ 'regular' if it is restricted and if Eq_ϕ is valid (i.e. true in any second-order model).

Lemma 12. Suppose that the L-criterion ϕ is regular. Then $Eq_=$ is true in any term domain T.

Proof. We consider the T-truth of $\forall xyz (x = y \ \& \ y = z \rightarrow x = z)$ (the other cases being similar). The proof is by induction on the rank α of T. If $\alpha = 0$, then T is empty and the result is immediate. So suppose $\alpha > 0$. Let s, t, and u be terms from T. Then we need to show that $s = t \ \& \ t = u \rightarrow s = u$ is true in T. So suppose that $s = t$ and $t = u$ are true (in T). If one of s or t or u is an individual constant, then all of them are by clause (ii) in the truth-definition. But then $s = t$ and $t = u$ by clause (i); so $s = u$; and so $s = u$ is true in T by clause (ii) again.

We may therefore suppose that all of s, t, and u are abstract terms. So put $s = \S C, t = \S D$ and $u = \S E$. Since $s = t$ is true (in T), $\phi(C, D)$ is true in $T_{C \cup D}$. But $rk(T_{C \cup D \cup E}) < rk(T)$ (since $rk(F) = rk(\S F) < rk(T)$ for any term $\S F$ of T); and so by IH, $Eq_=$

is true in $T_{C \cup D}$. By Lemma 10, $\phi[|C|, |D|]$ is true in $M_{C \cup D}$; so, by Lemma 11, $\phi[|C|_U, |D|_U]$, with $U = C \cup D \cup E$, is true in the restriction of M_U to $\{|t|_U : t \in C \cup D\}$; and so, by Lemma 3.12 and the fact that ϕ is restricted, $\phi[|C|_U, |D|_U]$ is true in M_U. Similarly, $\phi[|D|_U, |E|_U]$ is true in M_U. But then from the validity of Eq_ϕ, $\phi[|C|_U, |E|_U]$ is true in M_U. By the use of Lemma 11 and Lemma 3.12 again, $\phi[|C|_{C \cup E}, |E|_{C \cup E}]$ is true in $M_{C \cup E}$; by Lemma 10, $\phi(C, E)$ is true in the term domain $T_{C \cup E}$; and hence, by the truth-definition, $\S C = \S E$ is true in T.

How might a §-model be formed from a term domain? We cannot in general expect that § will be defined in M_T for every concept $|C|$ with $C \subseteq T$. The object term $\S T$ itself, for example, will not belong to T. Accordingly, let us say that the domain T is *representative* if for each object term s there is an object term t in T such that $s = t$ is true. Thus a representative domain represents the totality of objects that can be denoted by terms.

Given a representative domain T and a regular L-criterion ϕ, let M_T be a standard §-model, with domain M_T as before and with § defined, given any $C \subseteq T$, by:

$$\S(|C|) = |t|, \text{ for } t \in T \text{ and } \S C = t \text{ true.}$$

Each subset $C \subseteq M_T$ will be of the form $|C|$ for $C \subseteq T$; and hence § will, putatively, be defined on every concept of M_T. If the domain is representative, there will be a t conforming to the definiens for each $C \subseteq T$; and hence $\S$ will have a value for each argument $|C|$. Given that ϕ is regular, $Eq_=$ will be true in T by Lemma 12, and so the value $|t|$ will not depend upon the choice of t. Nor will the value depend upon the choice of C. For suppose $|C| = |D|$. Then $\phi[|C|, |D|]$ is true in the second order model M_T and hence, by ϕ restricted, is also true in the restriction $M_{C \cup D}$. Given that $Eq_=$ is true in T, it follows from Lemmas 10 and 11 that $\phi(C, D)$ is true in $T_{C \cup D}$ and hence that $\S C = \S D$ is true.

We can now extend Lemma 10 to the language $L^\S$:

Lemma 13. Suppose that ϕ is a regular L-criterion and that T is representative term domain. Let $\psi = \psi(x_1, \ldots, x_m, P_1, \ldots, P_n)$ be a formula of $L^\S$ with free variables as displayed; and let $t_1, \ldots, t_m$ be object terms of T and $Q_1, \ldots, Q_n$ be relation terms of the same arity, respectively, as $P_1, \ldots, P_n$. Then $\psi(t_1, \ldots, t_m, Q_1, \ldots, Q_n)$ is true in T iff $\psi[|t_1|, \ldots, |t_m|, |Q_1|, \ldots, |Q_n|]$ is true in M_T.

Proof. Building upon the inductive proof of Lemma 10, it suffices to show that the formula $\psi(C, D) = (\S C = \S D)$ is true (in T) iff $\psi[\S|C|, \S|D|]$ is true in M_T, i.e. iff $\S(|C|) = \S(|D|)$. But the latter holds iff for some t and u in T, $\S C = t, \S D = u$ are true and $|t| = |u|$, i.e. iff for some t in T, $\S C = t = \S D$. But given that T is representative, this holds iff $\S C = \S D$ is true.

We finally obtain our central result, that the truth- definition, relativized to a representative domain of terms, will yield the same truth-values as the corresponding minimal model.

Theorem 14. Suppose that ϕ is a regular L-criterion and that T is representative term domain. Then:

   (i) the corresponding §-model M_T is a minimal model of Φ; and
   (ii) a sentence ϕ of $L^\S$ is true in T iff it is true in M_T.

Proof. (i) may be straightforwardly established by using an induction on the rank of $t \in T$ in order to show that $|t|$ must be a member of the generated subdomain G of M_T. (ii) follows from Lemma 13.

The existence of a representative domain with **d** individual constants will imply the existence of a standard model for Φ (indeed, of a minimal model) with **d** individuals. The converse also holds. For if there exists a standard model with **d** individuals, it will have a minimal submodel with **d** individuals. Suppose now that α is the critical ordinal for the generating function G_ξ. Then it is readily shown that the set of terms of rank $\leq \alpha$ (or of rank $< \alpha$ for α a limit ordinal) will be a representative domain.

We might compare the above proof of soundness to the attempted demonstration of the soundness of Law V in sect. 31 of Frege's *Grundgesetze*. The proof there is infected by two sources of impredicativity: one to the predicates and the other to the abstraction operator (see Dummett 1991*a*: chap. 17). The first is removed here by adopting a Platonic conception of relations and concepts; the second is removed by assuming the existence of a representative domain of object terms. Whether such a domain exists depends, of course, on the identity criterion in question. In the case of coextensionality, there will be no such domain; while in the case of equinumerosity, there will.

Although my interest has principally been in a Platonic conception of concepts, it is worth pointing out it is also possible to obtain models of the abstraction theories in which all objects and all concepts are denoted by closed terms from the language itself (assuming that we allow the use of λ to form complex predicates). Thus from the results of Boolos (1968), we can obtain a term model for Hume's Law in which the concepts are taken, in effect, to be the predicatively defined sets of numbers. Also, as Tony Martin has pointed out, by assuming the axiom of projective determinacy, we can obtain term models that are elementary submodels of the standard model (conceived now in a first-order way); and it seems likely that these methods could be extended to various other forms of abstraction. We would thereby obtain something closer to the kind of model that Frege seemed to have in mind, though without the straightforward inductive determination of the truth-conditions.

7. Invariance

We shall attempt to characterize the non-inflationary invariant equivalence relations on a given power set. The analysis will then be extended to other forms of invariance.

Given an equivalence relation $\approx$ on the power set $\wp(M)$ of a set M, let $P_\approx$ be the partition of $\wp(M)$ induced by $\approx$ and call the equivalence relation $\approx$ *non-inflationary* if card$(P_\approx) \leq$ card(M). Each L-criterion ϕ induces an invariant relation $\approx$ on the powerset $\wp(M)$ of a set M of urelements; and ϕ will be tenable on the corresponding standard model M just in case $\approx$ is an equivalence relation and the induced partition $P_\approx$ is non-inflationary. Thus determining the non-inflationary equivalences on $\wp(M)$ will help us determine which L-criteria are tenable.

In the ensuing discussion, we take M to be a set of urelements; and, unless otherwise indicated, we shall assume that M is infinite. Recall that a *bicardinal* is an ordered pair $<\mathbf{c}, \mathbf{d}>$ of cardinals; and given a subset C of M, we set bicard$(C) =<$ card(C), card$(M - C)>$. Suppose that $\mathbf{m}$ is the cardinality of M. If card$(C) < \mathbf{m}$, then bicard$(C) =<$ card$(C), \mathbf{m}>$. Thus bicardinality, in this case, is a function of cardinality: card$(C) =$ card(D) implies bicard$(C) =$ bicard(D). On the other hand, if card$(C) = \mathbf{m}$, then bicard(C) can take any of the values $<\mathbf{m}, \mathbf{d}>$ for $\mathbf{d} \leq \mathbf{m}$.

Let C and D be any two subsets of M. Then the *cardinality distribution* cdstr(C,D) *of* C *and* D is the ordered quadruple $<$ card$(C - D)$, card$(C \cap D)$, card$(D - C)$, card$(M - (C \cup D))>$. Thus the cardinality distribution of two sets identifies the cardinalities of the smallest demarcated areas in their Venn diagram.

We first consider the question: when does identification of one pair of subsets of M under an invariant equivalence relation imply identification of another pair? Say that the pair $< C, D >$ of subsets of M *yields* the pair $< E, F >$ if, for any invariant equivalence $\approx$ over $\wp(M)$, $C \approx D$ implies $E \approx F$.

Cardinality distribution provides a sufficient condition for one pair of sets to yield another:

Lemma 1. $< C, D >$ yields $< E, F >$ if $\mathrm{cdstr}(C,D) = \mathrm{cdstr}(E,F)$.

Proof. Suppose that $\mathrm{cdstr}(C,D) = \mathrm{cdstr}(E,F)$, i.e. $\mathrm{card}(C \cap D) = \mathrm{card}(E \cap F)$, $\mathrm{card}(C-D) = \mathrm{card}(E-F)$, $\mathrm{card}(D-C) = \mathrm{card}(F-E)$, and $\mathrm{card}(M-(C \cup D)) = \mathrm{card}(M-(E \cup F))$. Let f_1 be a one-to-one map from $C \cap D$ onto $E \cap F$, f_2 a one-to-one map from $C-D$ onto $E-F$, f_3 a one-to-one map from $D-C$ onto $F-E$, and f_4 a one-to-one map from $M-(C \cup D)$ onto $M-(E \cup F)$). Then $f = f_1 \cup f_2 \cup f_3 \cup f_4$ is a permutation on M with $f[C] = E$ and $f[D] = F$. Suppose now that $C \approx D$. Given that $\approx$ is invariant, $f[C] \approx f[D]$, i.e. $E \approx F$.

Use of this lemma will often be implicit. Once it has been shown that $C \approx D$ for sets C and D of given cardinality distribution, it will be taken for granted that $E \approx F$ holds for any sets E and F of the same cardinality distribution.

By a *combination* is meant a pair $< C, D >$ of subsets of M of the same bicardinality. A combination $< C, D >$ of subsets of M (of bicardinality $< \mathbf{a}, \mathbf{b} >$) is said to be *representative* if it yields all combinations of subsets of M of the same bicardinality, i.e. for any invariant equivalence relation $\approx$ on $\wp(M)$, $C \approx D$ implies $E \approx F$ whenever $\mathrm{bicard}(E) = \mathrm{bicard}(F) = < \mathbf{a}, \mathbf{b} >$.

We wish to determine which combinations of sets are representative. The relation of *being almost the same* will be especially significant in this regard. Given two sets C and D, define their *symmetric difference* $C \sim D$ to be $(C-D) \cup (D-C)$. We say that two subsets C and D of M are *almost the same*—in symbols, $C \equiv' D$—if either (i) they are finite and $C = D$ or (ii) they are infinite, $\mathrm{card}(C) = \mathrm{card}(D)$ and $\mathrm{card}(C \sim D) < \mathrm{card}(C)$; and we say that the subsets C and D are *very different* if they are not almost the same.

Lemma 2. $\equiv'$ is an equivalence relation on the subsets of M.

Proof. The relation $\equiv'$ is clearly reflexive and symmetric. To establish transitivity, suppose that $C \equiv' D$ and $D \equiv' E$. We wish to show $C \equiv' E$. If one of C, D, or E is finite, then they are all finite and the result is obvious. So suppose that they are all infinite. Then $\text{card}(C) = \text{card}(D) = \text{card}(E)$. So it remains to show that $\text{card}(C-E) < \mathbf{c}$ and that $\text{card}(E-C) < \mathbf{c}$, where $\mathbf{c} = \text{card}(C)$. By symmetry, it suffices to establish the first of the two inequalities. Now $C-E \subseteq ((C-D) \cup (D-E))$. So $\text{card}(C-E) \leq \text{card}((C-D) \cup (D-E)) < \mathbf{c} + \mathbf{c} = \mathbf{c}$ (given that $\mathbf{c}$ is infinite).

A subset C of M is said to be *small* (*relative to M*) if $\text{card}(C) < \text{card}(M)$ and otherwise is said to be *large* (*relative to M*). (Later, another notion of smallness and largeness will be introduced.) When we are dealing with small subsets C and D of a set M of infinite cardinality $\mathbf{m}$, the cardinality of $M - (C \cup D)$ will always be $\mathbf{m}$ and hence may be ignored in assessing cardinality distribution.

We divide the results on representative combination into three parts:

Lemma 3 (part 1). A combination $< C, \ D >$ of small infinite sets is representative when they are very different.

Proof. Suppose $\text{card}(C) = \text{card}(D) = \mathbf{c} < \mathbf{m}$. We distinguish two subcases:

(1) C and D are disjoint. Take any two sets E and F of cardinality $\mathbf{c}$ (so that $\text{card}(M-E) = \text{card}(M-F) = \mathbf{m}$). Given that $C \approx D$ for invariant $\approx$, we wish to show that $E \approx F$. Clearly, we can pick a subset B of M of cardinality $\mathbf{c}$ disjoint from both E and F. (At this and similar points in the proofs the reader might like to draw a Venn diagram). Since $\text{cdstr}(C, \ D) = \text{cdstr}(E, \ B)$, it follows from Lemma 1 that $E \approx B$. Similarly, it follows from the fact that $\text{cdstr}(C, \ D) = \text{cdstr}(F, \ B)$ that $F \approx B$. But then $E \approx F$.

(2) C and D overlap. Since C and D are very different, either $C-D$ or $D-C$ is of cardinality $\mathbf{c}$. Without loss of generality, suppose that $C - D$ is. Let us also suppose that $\text{card}(C \cap D) = \mathbf{d}$ and that $\text{card}(D-C) = \mathbf{e}$. Pick a subset A of $C - D$ for which $\text{card}(A) = \mathbf{d}$ and $\text{card}(C-(D \cup A)) = \mathbf{e}$; and pick a subset B of $M-(C \cup D)$ of cardinality $\mathbf{e}$ (note that at least one of $\mathbf{d}$ or $\mathbf{e}$ is $\mathbf{c}$). Let $D^+ = A \cup B$. Then $\text{card}(D^+) = \mathbf{d} + \mathbf{e} = \mathbf{c}$, $\text{card}(C-D^+) = \mathbf{c}$, and $\text{card}(D^+ - C) = \mathbf{e}$. Thus $\text{cdstr}(C, \ D^+) = \text{cdstr}(C, \ D)$. So $C \approx D^+$ and, given $C \approx D$, $D^+ \approx D$. Now D^+ and D are disjoint. By (1) above, the combination $< D^+, \ D >$ is representative; and therefore so is the combination $< C, \ D >$.

Say that a subset C of the infinite set M is *almost universal* (*relative to M*) if card$(M - C) <$ card$(C) =$ card(M). Thus an almost universal set is one which is almost the same as the universal set M.

Lemma 3 (part 2). A combination $< C, \ D >$ of almost universal subsets of M is representative when their complements $M - C$ and $M - D$ are infinite and very different.

Proof. We reduce this case to the other. Given any equivalence relation $\approx$ on $\wp(M)$, we define the dual relation $\approx'$ by:

$C \approx' D$ iff $M - C \approx M - D$.

It is readily established that $\approx'$ is an equivalence relation and that it is invariant if $\approx$ is.

Now suppose that $C \approx D$ for $< C, \ D >$ a combination of almost universal sets. Then $M - C \approx' M - D$. Given that C and D are almost universal, $M - C$ and $M - D$ are small; and given that C and D are of the same bicardinality, so are $M - C$ and $M - D$. Since $M - C$ and $M - D$ are very different, it follows by the first part of the lemma that $C' \approx' D'$ for any sets C' and D' of the same bicardinality as $M - C$. But then $M - C' \approx M - D'$ for any such sets C' and D'; and so $E \approx F$ for any sets E and F of the same bicardinality as C.

Say that the set C *bifurcates* D if C is a subset of D and card$(C) =$ card$(D - C)$ and that the subset C of M is a *bifurcator* if it bifurcates M. A bifurcator divides the universe M into parts of equal size; and it will, for an infinite universe, be of the same size as the universe. From among the large subsets of M, the bifurcators are those that are not almost universal.

Lemma 3 (part 3). A combination $< C, \ D >$ of bifurcators C and D is representative when C is very different both from D and from $M - D$.

Proof. We distinguish three subcases:

(1) C and D are disjoint.

Proof. Clearly, $M - (C \cup D)$ must be large, for otherwise C would be almost the same as $M - D$. We first show that $E \approx F$ must hold for any bifurcators E and F whenever $M - (E \cup F)$ is large. For choose a bifurcator B of $M - (E \cup F)$. Then cdstr$(E, B) =$ cdstr(C, D) $=$ cdstr(F, B). So $E \approx B$ and $F \approx B$; and consequently, $E \approx F$. Now consider the case in which $M - (E \cup F)$ is small. Then both $E - F$ and

$F - E$ must be large. Choose a bifurcator B of $E - F$. Then all of $B \approx E$, $F \approx F - E$ and $B \approx F - E$ hold by what was shown earlier; and so $E \approx F$.

(2) $C \cap D$ is small.

Proof. In this case, $C - D$ and $D - C$ must both be large. But also $M - (C \cup D)$ must be large, for otherwise C would be almost the same as $M - D$. Suppose $\mathrm{card}(C \cap D) = \mathbf{c}$. Choose a subset A of $C - D$ of cardinality $\mathbf{c}$ and a bifurcator B of $M - (C \cup D)$. Let $D^+ = A \cup B$. Then it is readily verified that $\mathrm{cdstr}(C, D) = \mathrm{cdstr}(C, D^+)$. So $C \approx D^+$; and hence $D^+ \approx D$. But D and D^+ are disjoint. By case 1, the combination $< D, \ D^+ >$ is representative; and therefore so is the combination $< C, \ D >$.

(3) $C \cap D$ is large.

Proof. Split $C \cap D$ into two disjoint parts B_1 and B_2, each of cardinality $\mathbf{m}$. Since C and D are very different, one of $C - D$ or $D - C$ is large. Suppose, without loss of generality, that it is $C - D$. Let $D_1 = (D - C) \cup B_1$ and $D_2 = (D - C) \cup B_2$.

Then $\mathrm{cdstr}\,(C, D_1) = \mathrm{cdstr}(C, D)$ and $\mathrm{cdstr}(C, D_2) = \mathrm{cdstr}(C, D)$. Given $C \approx D$, $C \approx D_1$ and $C \approx D_2$; and hence $D_1 \approx D_2$. So we have a reduction to case (2) in which the intersection $(D_1 \cap D_2)$ is small.

It should be noted that, for bifurcators C and D, $C \equiv' D$ iff $M - C \equiv' M - D$; for the symmetric differences of the sets C and D and of their complements are the same and the cardinalities of the sets and their complements are also the same.

Summing up:

Theorem 4 (Representative Combinations). A combination $< C, D >$ of infinite subsets of M is representative if (i) C and D are small but very different, or (ii) C and D are almost universal but with infinite very different complements, or (iii) C and D are bifurcatory with one very different from the other and from its complement.

Another way to present our results (and others like them) is in terms of what one might call a calculus of cardinality distributions. Suppose that Γ is a set of possible cardinality distributions and that τ is a particular such distribution (for a given set M). We might say that Γ *entails* τ if any invariant equivalence on $\wp(M)$ that identifies sets C and D for which $\mathrm{cdstr}\,(C, D) \in \Gamma$ will identify sets C and D for which cdstr $(C, D) = \tau$; and we might then attempt to determine a sound and complete 'calculus' for the relation of entailment as so defined. However, this is not the approach we shall pursue.

We now attempt to take account of the non-inflationary character of an equivalence $\approx$. To this end, we need further information on the relation $\equiv'$ of approximate identity. For each subset C of M, let $|C|_{\equiv'} = \{D \subseteq M: C \equiv' D\}$, let $Q = \{|C|_{\equiv'}: C \subseteq M\}$, and let $Q_c = \{|C|_{\equiv'}: C \subseteq M \text{ and } C \text{ is of cardinality } \mathbf{c}\}$. Recall that $[M]^{\mathbf{c}}$ is the set of all subsets of M of cardinality $\mathbf{c}$. We may then show that the identification of approximately identical sets does not reduce their cardinality.

Lemma 5. card$(Q_c) = \mathbf{m}^{\mathbf{c}}$ for $\mathbf{m}$ transfinite and $\mathbf{c} \leq \mathbf{m}$.

Proof. For $\mathbf{m}$ transfinite and $\mathbf{c} \leq \mathbf{m}$, $\mathbf{m}^{\mathbf{c}}$ is the cardinality of $[M]^{\mathbf{c}}$; and since card$(Q_c) \leq$ card$([M]^{\mathbf{c}})$, card$(Q_c) \leq \mathbf{m}^{\mathbf{c}}$.

The other direction is proved by 'magnifying' the differences between distinct sets. To this end, take $\mathbf{c}$ disjoint copies M_ζ of M. Clearly, $M^+ = \bigcup \{M_\zeta: \zeta < \mathbf{c}\}$ is of cardinality $\mathbf{m}$. For each $\zeta < \mathbf{c}$, let f_ζ be a one-to-one map from M onto M_ζ. For each subset C of M of cardinality $\mathbf{c}$, let $f(C) = \bigcup \{f_\zeta[C]: \zeta < \mathbf{c}\}$. Thus f is a map from $[M]^{\mathbf{c}}$ into $[M^+]^{\mathbf{c}}$. Moreover, if $C \neq D$, then not $f(C) \equiv' f(D)$. For if $C - D$, let us say, contains an element a, then $f(C) - f(D)$ contains $\mathbf{c}$ copies of a and hence is of cardinality $\mathbf{c}$. So if $C \neq D$, then $|f(C)|_{\equiv'} \neq |f(D)|_{\equiv'}$(wrt M^+); and hence card$(Q_c) \geq$ card$([M]^{\mathbf{c}})$ $= \mathbf{m}^{\mathbf{c}}$.

Let us call an equivalence $\approx$ on $\wp(M)$ *strictly acceptable* if it is invariant and non-inflationary. Our characterization of the strictly acceptable equivalences will be in terms of a minimal such relation. A cardinal $\mathbf{c} \leq \mathbf{m}$ is said to be *exponentially small* (*relative to* $\mathbf{m}$) if $\mathbf{m}^{\mathbf{c}} \leq \mathbf{m}$ and *exponentially large* (*relative to* $\mathbf{m}$) if $\mathbf{m}^{\mathbf{c}} > \mathbf{m}$. The subset C of M is said to be *exponentially small* if card$\{D \subseteq M: \text{card}(D) \leq \text{card}(C)\} > \text{card}(M)$; and otherwise it is said to be *exponentially large*. If M is finite, then C is an exponentially small subset of M iff C is empty and M is non-empty. If M is infinite, then C is an exponentially small subset of M iff card(C) is exponentially small relative to card(M). We note that, by Cantor's Theorem, $\mathbf{m} > 1$ is itself exponentially large (relative to $\mathbf{m}$) and that if the generalized continuum hypothesis holds then every transfinite cardinal less than $\mathbf{m}$ is exponentially small (relative to $\mathbf{m}$). It is important to observe that a cardinal $\mathbf{c}$ may be exponentially small relative to a cardinal $\mathbf{m}$

and yet exponentially large relative to a larger cardinal $\mathbf{n}$. For example, given GCH, $\aleph_1^{\aleph_0} = \aleph_1$, whereas $\aleph_\omega^{\aleph_0} = \aleph_{\omega+1}$.

We define the *basal* relation $\approx_0$ on $\wp(M)$ by:

> $C \approx_0 D$ iff either (i) card(C) and card(D) are exponentially small and $C = D$, or (ii) card$(M - C)$ and card$(M - D)$ are exponentially small and $C = D$, or (iii) card(C), card(D), card$(M - C)$ and card$(M - D)$ are exponentially large and bicard$(C) = $ bicard(D).

Thus $\approx_0$ is a kind of hybrid relation—behaving like identity on sets that are sufficiently small or have sufficiently small complements and behaving like bi-equinumerosity on the remaining sets. By the basal relation *for* the cardinal $\mathbf{m}$, we mean the basal relation on $\wp(M)$ for M a set of urelements of cardinality $\mathbf{m}$. The basal relation for a cardinal is clearly unique up to isomorphism.

Say that the relation R is *as refined as* the relation S, or that S *subsumes* R, if S holds between any two objects between which R holds. We then have:

Theorem 6 (Characterization). Suppose that M is infinite. Then the basal relation $\approx_0$ is the most refined strictly acceptable equivalence relation on $\wp(M)$.

Proof. Let us first show that $\approx_0$ is a strictly acceptable equivalence. It is readily verified to be an equivalence and, from its characterization, is clearly invariant.

To show that $\approx_0$ is non-inflationary, let us count the equivalence classes $|C|$ in the partition P_0 induced by $\approx_0$. Consider first the C for which card$(C) = \mathbf{c}$ is exponentially small. There are $\mathbf{m^c}$ subsets C of M of cardinality $\mathbf{c}$ and hence $\mathbf{m^c}$ corresponding equivalence classes $|C|$ in P_0. But given that $\mathbf{c}$ is exponentially small, $\mathbf{m^c} \leq \mathbf{m}$. Consider now the C for which card$(M - C)$ is exponentially small. Then by similar reasoning, there are again $\mathbf{m^c}$ corresponding equivalence classes $|C|$ in P_0. Finally, consider the C for which both C and $M - C$ are exponentially large. Let $\mathbf{n}$ be the cardinality of the set of cardinals $\leq \mathbf{m}$. Then $\mathbf{n} \leq \mathbf{m}$ and there are at most $2 \cdot \mathbf{n} = \mathbf{n}$ such equivalence classes $|C|$ (one for each suitable bicardinality $<\mathbf{c}, \mathbf{d}>$, $\mathbf{c}, \mathbf{d} < \mathbf{m}$).

We see that in each case there are at most $\mathbf{m}$ equivalences classes. But there are at most $\mathbf{n}$ cases; and hence card$(P_0) \leq \mathbf{m}$.

Let us now show that $\approx_0$ is the most refined of the strictly acceptable equivalences. Consider an arbitrary non-inflationary and invariant equivalence $\approx$. Suppose that $C \approx_0 D$. We then wish to show $C \approx D$.

Let $\mathbf{c} = \text{card}(C)$ and $\mathbf{d} = \text{card}(M - C)$ (given $C \approx_0 D$, $\text{card}(C) = \text{card}(D)$ and $\text{card}(M - C) = \text{card}(M - D)$). If either $\mathbf{c}$ or $\mathbf{d}$ is exponentially small, it follows from the definition of $\approx_0$ that $C = D$; and so it is evident that $C \approx D$. So we may suppose that $\mathbf{c}$ and $\mathbf{d}$ are both exponentially large, i.e. $\mathbf{m^c} > \mathbf{m}$ and $\mathbf{m^d} > \mathbf{m}$. It then follows from the definition of $\approx_0$ that $\text{bicard}(C) = \text{bicard}(D)$.

Suppose, for reductio, that not $C \approx D$. By Theorem 4 on representative combinations (TRC for short), $\approx$ cannot hold between any representative pair of subsets of M of the same bicardinality as C and D. There are three cases, each of which will require $\approx$ to be inflationary.

Case 1. $\mathbf{c} < \mathbf{m}$. Then $\text{card}(M - C) = \text{card}(M - D) = \mathbf{m}$; C and D are infinite, since M is infinite and both C and D are exponentially large; and so, by part 1 of TRC, $E \approx F$ implies $E \equiv' F$ for any subsets of E and F of M of cardinality $\mathbf{c}$. But then:

$$\text{card}(P_\approx) \geq \text{card}(Q_c)$$
$$= \mathbf{m^c} \text{ (by Lemma 5)}$$
$$> \mathbf{m}, \text{ given that } \mathbf{c} \text{ is exponentially large.}$$

Case 2. $\mathbf{c} = \mathbf{m}$ and $\mathbf{d} < \mathbf{m}$. In a similar way to case 1, it follows by part 2 of TRC that $E \approx F$ implies $(M - E) \equiv' (M - F)$ for any subsets $M - E$ and $M - F$ of cardinality $\mathbf{d}$. But then:

$$\text{card}(P_\approx) \geq \text{card}(\{|(M - E)|_{\equiv'} : \text{card}(M - E) = \mathbf{d}\}$$
$$= \text{card}(Q_d)$$
$$= \mathbf{m^d} \text{ (by Lemma 5)}$$
$$> \mathbf{m}, \text{ given that } \mathbf{d} \text{ is exponentially large.}$$

Case 3 $\mathbf{c} = \mathbf{d} = \mathbf{m}$. By part 3 of TRC, $E \approx F$ implies that either $E \equiv' F$ or $E \equiv' (M - F)$ for bifurcators E and F. Define $\equiv''$ on bifurcators by: $E \equiv'' F$ iff $E \equiv' F$ or $E \equiv' (M - F)$. It is readily shown that $\equiv''$ is an equivalence relation. Define a map f from $P_{\equiv'}$ (as restricted to bifurcators) into $P_{\equiv''}$ by: $f(|C|_{\equiv'}) = |C|_{\equiv''}$. Then it should be clear that f maps exactly two equivalence classes from $P_{\equiv'}$ into each of the equivalence classes from $P_{\equiv''}$; and so $\text{card}(P_\approx) \geq \text{card}(P_{\equiv'}) = \text{card}(Q_m) = \mathbf{m^m} > \mathbf{m}$.

We now consider the case in which the equivalence relation is required to be internally invariant. We say that an equivalence on $\wp(M)$ is *strictly I-acceptable* if it is I-invariant and non-inflationary. Define the *super-basal* relation $\approx_1$ on $\wp(M)$ by:

$C \approx_1 D$ iff either (i) card(C) and card(D) are exponentially small and $C = D$ or (ii) card(C) and card(D) are exponentially large and card(C) = card(D).

We then have:

Corollary 7. Suppose that M is infinite. Then the relation $\approx_1$ is the most refined strictly I-acceptable equivalence on $\wp(M)$.

Proof. Given the theorem, it suffices to show that $\approx_1$ is the smallest I-invariant equivalence to contain $\approx_0$. It is readily verified that $\approx_1$ is an I-invariant equivalence. Now consider any I-invariant equivalence $\approx$ containing $\approx_0$; and suppose that $C \approx_1 D$. There are two cases: (i) card(C) and card(D) are exponentially small and $C = D$. But then it is evident that $C \approx D$. (ii) card(C) and card(D) are exponentially large and card(C) = card(D). Suppose that card($C \cap D$) = $\mathbf{e}$. Clearly, we can find subsets C' and D' of M such that card(C') = card(D') = card(C), card($C' \cap D'$) = $\mathbf{e}$, and card($M - C'$) = card($M - D'$) = $\mathbf{m}$. There is therefore a one-to-one map f from $C \cup D$ onto $C' \cup D'$ with $f[C] = C'$ and $f[D] = D'$. So by the internal invariance of $\approx_1$, $C' \approx_1 D'$. But card(C') (= card(D')) and card($M - C'$) (= card($M - D'$)) are both exponentially large; and so $C' \approx_0 D'$. Given that $\approx$ contains $\approx_0$, $C' \approx D'$; and so by the internal invariance of $\approx$, $C \approx D$.

According to the theorem, the most refined form of invariant abstraction, compatible with full second-order logic, is one that yields the extensions of concepts whose extensions or counter-extensions are exponentially small and yields the bicardinalities of the concepts otherwise. According to the corollary, if it is further required that the identity criterion be absolute, then the most refined form of abstraction is one that yields extensions of concepts whose extensions are exponentially small and yields the cardinalities of the concepts otherwise. Thus bi-equinumerosity and equinumerosity represent the 'outer limits' of invariant abstraction; any acceptable method of invariant abstraction must eventually degenerate into one of the equinumerosity relations or something coarser.

There are various strengthenings of our basic results, some of which will later prove useful. We concentrate on the case in which invariance is assumed to be internal, although related results can be established for the cases in which it is not. Suppose that M is a set of urelements of (finite or infinite) cardinality $\mathbf{m}$. For $\mathbf{c} \leq \mathbf{m}$ and

$\mathbf{d} \leq \mathbf{c}^+$ (the successor of $\mathbf{c}$), define the relation $\equiv_{\mathbf{d}}$ on the members of $[M]^{\mathbf{c}}$ by: $C \equiv_{\mathbf{d}} D$ if $\mathrm{card}(C \sim D) < \mathbf{d}$. For $\mathbf{d} = \mathbf{c}$, $\equiv_{\mathbf{d}}$ is coincident with the earlier relation $\equiv'$ of being almost the same on infinite sets of cardinality $\mathbf{c}$; and for $\mathbf{d} = (2.\mathbf{c}^+)$, $\equiv_{\mathbf{d}}$ identifies all the sets of cardinality $\mathbf{c}$. In the same way as for Lemma 2, it may be shown that the relation $\equiv_{\mathbf{d}}$ is an equivalence for transfinite $\mathbf{d}$.

It will be helpful to classify the consequences of $C \approx D$ holding in relationship to the cardinality of $C \sim D$. To this end, define the *internal cardinality distribution* Icdstr (C, D) *of* two subsets C and D of M by $< \mathrm{card}(C - D),\ \mathrm{card}(C \cap D),\ \mathrm{card}(D - C) >$ (the cardinality of $M - (C \cup D)$ is ignored). We can then obtain the analogue of Lemma 1 (using I-invariant equivalences in place of invariant equivalences).

When the cardinality $C \sim D$ is infinite, we have:

Lemma 8. Let $\approx$ be an I-invariant equivalence over the infinite set M; and suppose that $C \approx D$ holds with $\mathrm{card}(C \sim D) = \mathbf{d}$ infinite and $\mathrm{card}(C) = \mathrm{card}(D) = \mathbf{c}$. Then $E \equiv_{\mathbf{d}+} F$ implies $E \approx F$ for any sets E and F of cardinality $\mathbf{c}$.

Proof. Let us first consider the case in which $\mathbf{d} = \mathbf{c}$ (and C is consequently very different from D). We may then use part 1 of Lemma 3 to show that $E \approx F$ for any subsets E and F of M cardinality $\mathbf{c}$. For extend the domain M to a domain M^+ of cardinality $\mathbf{m}$ relative to which the particular sets C and D are small. Define $\approx^+$ by: $E \approx^+ F$ iff $\mathrm{Icdstr}(E, F) = \mathrm{Icdstr}(A, B)$ for some subsets A and B of M for which $A \approx B$. It is then readily shown that $\approx^+$ is an I-invariant equivalence on M^+. Pick now any subsets E and F of M of cardinality $\mathbf{c}$. Then $\mathrm{Icdstr}(E, F) = \mathrm{Icdstr}(A, B)$ for some subsets A and B of M for which $A \approx B$; and hence $E \approx^+ F$.

We may therefore assume that $\mathbf{d} < \mathbf{c}$. We may also assume that $\mathrm{card}(M - (C \cup D)) = \mathrm{card}(M)$. For C and D may be simultaneously mapped one-to-one onto subsets C' and D' for which $\mathrm{card}(M - (C' \cup D')) = \mathrm{card}(M)$; since $\mathrm{Icdstr}(C, D) = \mathrm{Icdstr}(C', D')$, $C' \approx D'$; and so it suffices to establish the result for subsets of this sort.

Let $\mathrm{card}(C - D) = \mathbf{d}_1$ and $\mathrm{card}(D - C) = \mathbf{d}_2$. The result will then follow from the following two facts:

(1) If $\mathbf{d} = \mathbf{d}_1 > \mathbf{d}_2$, then $E \approx F$ whenever $\mathrm{Icdstr}(E, F) = < \mathbf{d}_1, \mathbf{c}, \mathbf{d}_1 >$ or $\mathrm{Icdstr}(E, F) = < \mathbf{d}_2, \mathbf{c}, \mathbf{d}_2 >$;

(2) If $\mathbf{d}_1 = \mathbf{d}_2 = \mathbf{d}$, then $E \approx F$ whenever $\mathrm{Icdstr}(E, F) = < \mathbf{d}, \mathbf{c}, \mathbf{e} >$ for any $\mathbf{e} \leq \mathbf{d}$.

For suppose that $E \equiv_{\mathbf{d}+} F$ and $\mathrm{Icdstr}(E, F) = <\mathbf{f}, \mathbf{g}, \mathbf{h}>$. Then $\mathbf{f}, \mathbf{h} \leq \mathbf{d} < \mathbf{c}$; and so $\mathbf{g} = \mathbf{c}$. Assume, without loss of generality, that $\mathbf{f} \leq \mathbf{g}$ and suppose, in the first instance, that the subsets C and D are of a type for which $\mathbf{d}_1 = \mathbf{d}_2 = \mathbf{d}$. By (2), the result holds for any subsets E and F of internal cardinality distribution $<\mathbf{d}, \mathbf{c}, \mathbf{h}>$; and so, since there are subsets of this sort, it follows by (1) that the result holds for any subsets of internal cardinality distribution $<\mathbf{h}, \mathbf{c}, \mathbf{h}>$. If $\mathbf{f} = \mathbf{h}$, we are done; and if $\mathbf{f} < \mathbf{h}$, we may reapply (2). On the other hand, if the subsets C and D are of a type for which $\mathbf{d} = \mathbf{d}_1 > \mathbf{d}$, then we may apply (1) to obtain a case for which $\mathbf{d}_1 = \mathbf{d}_2 = \mathbf{d}$ and thereby effect a reduction to the first instance.

(1) and (2) may be proved as follows.

Proof of (1). Pick a subset B of $M - (C \cup D)$ of cardinality $\mathbf{d}$. Let $C' = (C \cap D) \cup B$. Then by considerations of cardinality distribution (CCD, for short), $C' \approx D$. Hence $C \approx C'$. But $\mathrm{Icdstr}(C, C') = <\mathbf{d}, \mathbf{c}, \mathbf{d}>$. The proof for the other case is similar.

Proof of (2). Pick a subset B of $M - (C \cup D)$ of cardinality $\mathbf{e} \leq \mathbf{d}$ and divide $C - D$ into two subsets C_1 and C_2 of cardinality $\mathbf{d}$. Let $C' = (C \cap D) \cup B \cup C_1$. By CCD, $C' \approx D$. Hence $C \approx C'$. But $\mathrm{Icdstr}(C, C') = <\mathbf{d}, \mathbf{c}, \mathbf{e}>$.

When the cardinality of $C \sim D$ is finite, we have:

Lemma 9. Let $\approx$ be an I-invariant equivalence on $\wp(M)$ (for M either finite or infinite). Suppose that $C \approx D$ holds for $\mathrm{card}(C \sim D)$ finite but non-zero and $\mathrm{card}(C) = \mathbf{c}$. Then $E \equiv_\infty F$ implies $E \approx F$ for any subsets E and F of M of cardinality $\mathbf{c}$.

Proof. Suppose that $\approx$ is an I-invariant equivalence on $\wp(M)$; and, for the purposes of the proof, let us define Ey/x by $(E - \{x\}) \cup \{y\}$. If C is empty, the result is trivial; and so we may assume that C is non-empty. In case M is finite, we may assume that $M\text{--}C$ is non-empty, since otherwise the result is again trivial; and in case M is infinite, we may assume that $\mathrm{card}(M - (C \cup D)) = \mathrm{card}(M)$ for the same reasons as before. In each of these cases, it then suffices to show that $C' \approx C$ for any subset C' of the same cardinality as C.

We distinguish three cases:

(i) D is a proper subset of C. Suppose that $x \in C - D$. Let y be a member of $M - C$; and set $D' = Cy/x$. By CCD

$D' \approx D$. But then $C \approx D'$; and so we have a reduction to case (iii) below.

(ii) C is a proper subset of D. Suppose that $x \in C$ and that $y \in D - C$. Let $D' = Cy/x$. By CCD, $D' \approx D$. But then $C \approx D'$; and we again have a reduction to case (iii) below.

(iii) C and D properly overlap. Suppose $z \in D - C$. Let n = card $(C \sim C')$, where C' is a subset of M of the same cardinality as C. We prove $C \approx C'$ by induction on n (since M may be finite, we need to take some special care in the reasoning). For n = 0, the result is trivial. So suppose the result holds for n ≤ k; and let us establish it for n = k + 1. Pick an element x in $C - C'$ and an element y in $C' - C$. Let $C^* = Cy/x$. It then suffices to show $C^* \approx C$: for card$(C^* \sim C') = (k - 1)$ and C^* and C properly overlap; so given $C^* \approx C$, it follows by IH that $C^* \approx C'$; and hence $C \approx C'$.

To show $C \approx C^*$, we distinguish three subcases:

(a) $y \in M - D$. Let $D' = Dy \backslash z$. Then by CCD, $C \approx D'$ and we have a reduction to case (b) or (c) below.

(b) $y \in D - C$ and $x \in M - D$. Suppose $u \in C \cap D$. Let $D' = Dx/u$. By CCD, $C \approx D'$ and we have a reduction to case (c) below.

(c) $y \in D - C$ and $x \in C \cap D$. But then $C^* = Cy/x \approx D$ by CCD; and so $C \approx C^*$.

As a special case of the lemma, note the following:

Corollary 10. Let $\approx$ be an I-invariant equivalence on $\wp(M)$. Suppose that $C \approx D$ holds for distinct finite sets C and D. Then $E \approx F$ holds for any subsets E and F of the same cardinality as C.

We consider finally the identification of sets of different cardinality. Say that an equivalence $\approx$ over M is *numeric on* the cardinal **c** if $C \approx D$ for any subsets C and D of cardinality **c**; and say that it is is *numeric simpliciter* if it is numeric on every cardinal ≤ card(M).

Lemma 11. Suppose that $\approx$ is an I-invariant equivalence over an infinite set M. If $C \approx D$ for sets C and D of distinct cardinality, then $\approx$ is numeric on the cardinality of C.

Proof. For C and D finite, the result follows from Corollary 10. So suppose that card$(C) <$ card(D) with D infinite. We may then find C' and D' such that Icdstr$(C', D') = $ Icdstr(C, D) and yet $M - (C' \cup D')$ is of the same cardinality as M. We may therefore find a C^* such that Icdstr$(C^*, D') = $ Icdstr(C', D') and C' is disjoint from C^*. So by

Theorem 4, $<C, C^*>$ is a representative combination, and the required result follows.

Now suppose that $\text{card}(D) < \text{card}(C)$ with C infinite. By the same reasoning as before, we may suppose that $M - (C \cup D)$ is of the same cardinality as M. Choose a subset C' of $M - (C \cup D)$ which is of the same cardinality as C and for which $M - (C \cup D \cup C')$ is of the same cardinality as M. Then we may find a D' such that $\text{Icdstr}(C, D) = \text{Icdstr}(C', D')$. By the previous case, $D \approx D'$; by the invariance of $\approx$, $C' \approx D'$; and so $C \approx C'$. So again by Theorem 4, $<C, C'>$ is a representative combination; and the required result follows.

Summing up Lemmas 8, 9, and 11, we obtain:

Theorem 12. Let $\approx$ be an I-invariant equivalence over the set M; and suppose that $C \approx D$ holds with $\text{card}(C \sim D)$ non-zero and $\text{card}(C) = \mathbf{c}$. Then $E \equiv_{\mathbf{d}} F$ implies $E \approx F$ for any sets E and F of cardinality $\mathbf{c}$, where $\mathbf{d} = \max\{\aleph_0, \text{card}(C \sim D)^+\}$.

The theorem provides a good picture of the structure of I-invariant equivalences. For subsets of a given finite cardinality, either no two sets of that cardinality are identified or all are; for sets of a given infinite cardinality $\mathbf{c}$, the sets are identified according to the relationship $\equiv_{\mathbf{d}}$ for some cardinal $\mathbf{d}$ ($\mathbf{d}$ will be the successor of $\sup\{\mathbf{e}: C \approx D, \text{card}(C) = \text{card}(D) = \mathbf{c}$, and $\text{card}(C \sim D) = \mathbf{e}\}$); and if two sets of distinct cardinality are identified, then all sets of those cardinalities are identified.

The lemmas have some further consequences of interest. From Lemmas 8, 10, and 11, we readily obtain:

Corollary 13. Suppose that $\approx$ is an I-invariant equivalence over infinite M. If $C \approx D$ for very different sets C and D, then $\approx$ is numeric on $\text{card}(C)$.

If we insist that the equivalence should be non-inflationary, then we also obtain:

Theorem 14. Suppose that $\approx$ is a non-inflationary I-invariant equivalence on $\wp(M)$, with M finite. Then $\approx$ is numeric.

Proof. Suppose that $C \approx D$ fails to hold for C and D of the same cardinality k. Then C and D must be distinct and non-empty; and so $0 < k < m = \text{card}(M)$. But then by Corollary 10, no two subsets of M of cardinality k can be related by $\approx$. Now the number of such subsets is at least m; and so, given that $\approx$ is non-inflationary, $M \approx C$ for some such set C. But then, by CCD, $M \approx C$ for every such set C; and the different Cs are related by $\approx$ after all.

Thus we see that the possibilities for non-inflationary identification are much more limited in the finite than in the infinite case. Any strictly I-acceptable equivalence must contain the relation of equinumerosity. But this relation is itself inflationary, of course, on finite domains, since for a domain M of n elements the relation yields $n + 1$ equivalence classes. Indeed, it is easy to see that there can be no smallest non-inflationary equivalence on $\wp(M)$ should $m = card(M) > 1$, since we are free to distinguish between the k-membered and the $(k + 1)$-membered subsets for *any* k = 0, $1, \ldots, (m - 1)$, but we are not free to distinguish between them for *all* k.

With each numeric equivalence $\approx$ on $\wp(M)$, we may associate an equivalence $\approx_{num}$ on all cardinals according to the rule:

> $\mathbf{c} \approx_{num} \mathbf{d}$ iff either (i) for some subsets C and D of M, $C \approx D$, $card(C) = \mathbf{c}$ and $card(D) = \mathbf{d}$ or (ii) $\mathbf{c}$, $\mathbf{d} > card(M)$.

Thus $C \approx D$ iff C and D are subsets of M and $card(C) \approx_{num} card(D)$. The equivalence classes induced by $\approx_{num}$ may be regarded as a form of generalized number.

There is a rather different way in which we may extend our basic results; for we need not require that the underlying relation be an equivalence. Call the relation $\leftrightarrow$ on an arbitrary set X a *biequivalence* if $x \leftrightarrow y$ & $x' \leftrightarrow y$ & $x' \leftrightarrow y'$ implies $x \leftrightarrow y'$. Given a biequivalence $\leftrightarrow$ on a set X, we say $x, y \leftrightarrow$ (x and y are *connected on the left*) if $x \leftrightarrow z$ and $y \leftrightarrow z$ for some z; and, similarly, we say $\leftrightarrow x, y$ (x and y are *connected on the right*) if $z \leftrightarrow x$ and $z \leftrightarrow y$ for some z. We also denote these two relations by $\leftrightarrow_L$ and $\leftrightarrow_R$. It is readily shown that the relation $\leftrightarrow_L$ is an equivalence relation on the domain of $\leftrightarrow$ and that $\leftrightarrow_R$ is an equivalence relation on the range of $\leftrightarrow$. Let us use $|x|_L$ and $|y|_R$ for the corresponding equivalence classes, and P_L and P_R for the corresponding partitions. For $X \in P_L$ and $Y \in P_R$, say that $X \leftrightarrow Y$ if $x \leftrightarrow y$ for some $x \in X$ and some $y \in Y$. Then it is readily shown that $\leftrightarrow$ is a one-to-one relation between P_L and P_R and that $x \leftrightarrow y$ iff $|x|_L \leftrightarrow |y|_R$ for any x in the domain D_L of $\leftrightarrow$ and any y in the range D_R of $\leftrightarrow$. Conversely, given a one-to-one relation $\leftrightarrow$ between partitions P_L and P_R of $\wp(D_L)$ and $\wp(D_R)$ respectively, we may determine the relation $\leftrightarrow$ between D_L and D_R to which it corresponds by means of the definition: $x \leftrightarrow y$ if $|x|_L \leftrightarrow |y|_R$.

Our interest is in the case in which $\leftrightarrow$ is a relation on the powerset $\wp(M)$. For this case, we have the following analogue of Theorem 12:

Theorem 15. Let $\leftrightarrow$ be an I-invariant biequivalence on $\wp(M)$; and suppose that $C \leftrightarrow D$ holds with $\text{card}(C \sim D)$ non-zero and $\text{card}(C) = \mathbf{c}$. Then $E \equiv_{\mathbf{d}} F$ implies $E \approx F$ for any sets E and F of cardinality c, where $\mathbf{d} = \max\{\aleph_0, \text{card}(C \sim D)^+\}$.

Proof. By a straightforward modification of the proof of Theorem 12. Details are omitted.

It is also possible to extend the above results to relations that are *relatively* invariant, i.e. to relations that ignore the identity of all but a fixed set of individuals. Say that a function f from X into Y is *fixed on* a subset X' of X if for all $x \in X'$, $f(x) = x$. For R a relation on $\wp(M)$ and K a subset of M, we then say that R is *K-invariant* if $R(C, D)$ implies $R(f(C), f(D))$ for any permutation f of M that is fixed on K and we say that R is *internally K-invariant* if $R(C, D)$ implies $R(f(C), f(D))$ for any one-to-one map from f from $C \cup D$ into M that is fixed on $K \cap (C \cup D)$.

For M an infinite set and K a subset of M, define the equivalence $\approx_{0,K}$ on $\wp(M)$ by:

$C \approx_{0,K} D$ iff $C \cap K = D \cap K$ and $C - K \approx_0 D - K$.

And similarly for $\approx_{1,K}$. Let us say that an equivalence $\approx$ on $\wp(M)$ is *(internally) K-acceptable* if it (internally) K-invariant and non-inflationary. We may then show:

Corollary 16. For M an infinite set and K an exponentially small subset of M, $\approx_{0,K}$ (resp. $\approx_{1,K}$) is the most refined (internally) K-acceptable equivalence on $\wp(M)$.

Proof. We prove the result for the relation $\approx_{0,K}$, the proof for the relation $\approx_{0,K}$, being exactly similar.

Given that $\approx_0$ is an equivalence relation, it is readily shown that $\approx_{0,K}$ is an equivalence relation; and it is evident that $\approx_{0,K}$ is internally K-invariant. Let us use the unadorned $|C|$ for the equivalence classes wrt $\approx_{0,K}$. For L a subset of M, let $M_L = \{C \subseteq M: C \cap K = L\}$. Then, for each $C, D \in M_L$, $C \approx_{0,K} D$ iff $C - K \approx_0 D - K$. So, by Theorem 6, the number of equivalence classes $|C|$ for $C \in M_L$ is at most $\mathbf{m} = \text{card}(M)$. But there are at most $2^{\mathbf{k}}$ subsets of K, where $\mathbf{k} = \text{card}(K)$. So the total number of equivalence classes $|C|$ is at most

$\mathbf{m}.2^{\mathbf{k}} \leq \mathbf{m}.\mathbf{m}^{\mathbf{k}} \leq \mathbf{m}$, given that $\mathbf{k}$ is exponentially small; and hence $\approx_{0,K}$ is non-inflationary.

To show that $\approx_{0,K}$ is the most refined K-acceptable equivalence, let $\approx$ be an arbitrary K-acceptable equivalence on M. For each subset L of K, define $\approx_L$ on $\wp(M-K)$ by:

$$C \approx_L D \text{ iff } C \cup L \approx D \cup L.$$

Then $\approx_L$ is a strictly acceptable equivalence on $\wp(M-K)$. So by Theorem 6, it contains $\approx_0$ as defined on $\wp(M-K)$ and hence is readily shown to contain the restriction of $\approx_0$, as defined on $\wp(M)$ to $\wp(M-K)$. Now suppose $C \approx_{0,K} D$ for $C, D \subseteq M$. Then $C-K \approx_0 D-K$ and $C \cap K = D \cap K$. Hence $C-K \approx_{C \cap K} D-K$; and so $C = (C-K) \cup (C \cap K) \approx (D-K) \cup (D \cap K) = D$.

It should be noted that the various minimal relations that we have appealed to can all be expressed within second-order logic. In order to express the absolute relations $\approx_0$ and $\approx_1$, we use the definitions:

Exponential Smallness: Expsmall(C) for $\exists R \forall D (D \leq C \rightarrow \exists y \forall x (xRy \leftrightarrow Dx))$ (the 'y' in the range of R here serves to enumerate those sets of lesser cardinality than C);

Exponential Largeness: Explarge(C) for $\neg$ Expsmall(C);

Suitability: Stb(C) for (Expsmall(C) $\vee$ $\exists D$ (D compl C & Expsmall(D)));

Unsuitability: Unstb(C) for $\neg$Stb(C).

The basal relation $\approx_0$ is then expressed by the formula $\phi_0(C, D)$:

(Stb(C) & Stb(D) & C $\equiv$ D) $\vee$ (Unstb(C) & Unstb(D) & C beq D).

Thus if M is an infinite set and $\approx_0$ is the basal relation on $\wp(M)$ then, for any subsets C and D of M, $C \approx_0 D$ iff $\phi_0[C, D]$ is true in the standard model $\mathbf{M}$ based on M. Similarly, the super-basal relation $\approx_1$ is expressed by the formula $\phi_1(C,D)$:

(Expsmall(C) & Expsmall(D) & C $\equiv$ D) $\vee$ (Explarge(C) & Explarge(D) & C eq D).

The relativized relations $\approx_{0,K}$ and $\approx_{1,K}$ are then expressed by relativizing the formulas ϕ_0 and ϕ_1 to a predicate K in the obvious way.

Using the above results, we can provide a deeper analysis of the various forms of acceptability. We limit our attention to the case of invariant (as opposed to K-invariant) criteria.

Theorem 17. Suppose that M is an infinite standard model. Then:

(i) An (absolute) L-criterion ϕ is tenable on M iff $E_{\phi, M}$ is an equivalence which subsumes the basal (resp. the super-basal) equivalence on $\wp(M)$;

(ii) An L-criterion ϕ is stable if, for each sufficiently large M, $E_{\phi, M}$ is an equivalence that subsumes the basal equivalence on $\wp(M)$;

(ii)' An absolute L-criterion ϕ is stable iff the corresponding global relation E_ϕ is an equivalence that subsumes equinumerosity for all sets of sufficiently large cardinality;

(iii) An L-criterion ϕ is generally tenable iff $E_{\phi, M}$ is an equivalence that subsumes the relation $\approx_{be}$ of bi-equinumerosity on $\wp(M)$ for each infinite model M;

(iii)' An absolute L-criterion ϕ is generally tenable iff the corresponding global relation $E_{\phi, M}$ is an equivalence that subsumes equinumerosity on $\wp(M)$ for each infinite set M of urelements.

Proof. (i) Since ϕ is an L-criterion, $E_{\phi, M}$ is invariant by Lemma 3.10. Now ϕ is tenable over M iff $E_{\phi, M}$ is an equivalence and non-inflationary, by Corollary 4.2. But given that $E_{\phi, M}$ is invariant, it follows from Theorem 6 above that it is a non-inflationary equivalence iff it contains the basal relation on $\wp(M)$.

The other results are proved similarly.

From (i) it follows that if the criterion $\phi = \phi(C, D)$ is tenable on M then the formula $\phi_0(C, D) \rightarrow \phi(C, D)$ will be true in the standard model M and, should ϕ be absolute, the formula $\phi_1(C, D) \rightarrow \phi(C, D)$ will also be true in M. But note that the criterion $\phi_0(C, D)$ is not itself absolute.

There is a most refined (global) relation that is generally tenable. It is defined, for sets C and D of urelements, by:

$C \approx_2 D$ iff (i) C and D are both finite and $C = D$, or (ii) C and D are both infinite and $\text{card}(C) = \text{card}(D)$.

This relation can be expressed by an absolute (indeed, by a restricted) formula. For let $\text{Fin}(C)$ be a restricted expression of finitude. Then $\approx_2$ can be expressed by the formula $\phi_2 = \phi_2(C, D)$:

$(\text{Fin}(C) \ \& \ \text{Fin}(D) \ \& \ C \equiv D) \vee (\neg\text{Fin}(C) \ \& \ \neg\text{Fin}(D) \ \& \ C \text{ eq } D)$.

Thus if we are to remain neutral about the size of the universe (apart from its being infinite), then ϕ_2 represents the best we can do using absolute means of expression. If we also insist on a 'uniform meaning' from one domain to another, then it would appear to be equinumerosity, *simpliciter*, that represents the best we can do; and this fact may go some way towards explaining the privileged role of Hume's Law in discussions of abstraction.

Our results can be extended to restricted theories of extensional abstraction, i.e. theories of the form $T^{\phi, \psi}$ where ϕ is the criterion $C \equiv D$.

Corollary 18. Suppose that $T = T^{\phi, \psi}$ is a restricted theory of extensional abstraction, with ψ logical, and that M is an infinite standard §-free model. Then some expansion of M is a model of T iff $E_{\phi, M}$ is an equivalence on $E_{\psi, M}$ and $E_{\psi, M}$ contains only exponentially small subsets of M.

Proof. Associated with the restricted principle $\Phi_\psi = [\psi(C) \& \psi(D) \rightarrow (\S C = \S D \leftrightarrow C \equiv D)]$ is the unrestricted principle Φ': $[\S C = \S D \leftrightarrow ((\psi(C) \& \psi(D) \& C \equiv D) \vee (\neg\psi(C) \& \neg\psi(D)))$. Thus in the associated unrestricted principle, all concepts that fail to conform to the given restriction are identified. It is then easy to show that some expansion of M is a model for Φ_ψ iff ϕ' is tenable on M. By Theorem 17(i) above, ϕ' is tenable on M iff $E_{\phi', M}$ is an equivalence that subsumes the basal equivalence $\approx_0$ on $\wp(M)$. But $E_{\phi', M}$ is an equivalence iff $E_{\phi, M}$ is an equivalence on $E_{\psi, M}$; and $E_{\phi', M}$ subsumes $\approx_0$ iff $E_{\psi, M}$ contains only exponentially small subsets of M.

Thus the limits of extensional abstraction are set by concepts with exponentially small extensions. When the domain M is countably infinite, the exponentially small sets are, of course, the same as those that are small in the usual sense of having a cardinality less than that of the universe. But for larger domains M, the two notions may diverge—some small sets in the usual sense may not be exponentially small.

Similar results can be established for abstraction theories T^Φ with several abstraction operators. For example, in analogy to Theorem 17 (i), we have:

Corollary 19. If M is an infinite standard model and T^Φ is a generalized abstraction theory, containing only logical abstraction principles, then some §-expansion of M is a model for T^Φ iff $E_{\phi, M}$ is an equivalence containing the basal equivalence on $\wp(M)$ for each of the identity criteria $\phi = \phi_\S$.

Proof. We may show that some §-expansion of *M* is a model of the generalized theory T^Φ iff for each § ϵ **§**, some §-expansion of *M* is a model of the individual theory T^Φ with $\phi = \phi_§$. The left-to-right direction is trivial; and for the right-to-left direction, we may simply 'piece together' all the component §-expansions to obtain the §-expansion. The result then follows from Theorem 17(i).

It should be noted that there is no guarantee that the §-expansion of *M* obtained under this corollary will be separated; for, in piecing together the component §-expansions, different abstracts may be identified. In the next section, we consider the possibilities for obtaining separated models.

8. Hyperinflation

We build on the previous analysis of invariance to determine the conditions under which there will be no hyperinflation among the non-inflationary invariant principles. The analysis is then extended to other forms of invariance.

By a *cell* is meant a non-empty collection of sets. A cell *X* is said to be *over* a set *M* if each member of *X* is a subset of *M*. A cell *X* is said to be *strictly admissible in M* if it is over *M* and if it is a member of $P_\approx$ for some strictly acceptable equivalence $\approx$ on $\wp(M)$. Let C_M (or simply **C** if *M* is understood) be the set of strictly admissible cells in *M*; and let $E_M = \{\approx : \approx$ is a strictly acceptable equivalence on $\wp(M)\}$. Our aim is to determine when card(C_M) and card(E_M) are $\leq$ card(*M*).

For $\approx$ an equivalence on $\wp(M)$, let cdstr($\approx$) = {cdstr(*C, D*): $C \approx D$}.

Lemma 1. $\approx$ and $\approx'$ are the same invariant equivalences on $\wp(M)$ if cdstr($\approx$) = cdstr($\approx'$).

Proof. Assume that the condition holds; and suppose that $C \approx D$. Then cdstr(*C, D*) ϵ cdstr($\approx$) and hence cdstr(*C, D*) ϵ cdstr($\approx'$). So for some C' and D', $C' \approx' D'$ and cdstr(*C', D'*) = cdstr(*C, D*). But then for some permutation *p* on *M*, $p[C'] = C$ and $p[D'] = D$; and so, by $\approx'$ invariant, $C \approx' D$.

For each cardinal **c**, let cp(**c**) = card({**d**: **d** a cardinal $\leq$ **c**}) 'cp' here stands for cardinality of predecessors and is meant to be reminiscent of 'cf' for cofinality. The notion will play an important role in our consideration of hyperinflation; and it is therefore of interest that the related finitary notion plays an important role in Frege's proof of the infinity of the natural numbers.

Lemma 2. For $\mathbf{m} = \text{card}(M)$ transfinite:
  (i) $\text{card}(E_M) = 2^{\text{cp}(\mathbf{m})}$;
  (ii) $\text{card}(C_M) = \max(\mathbf{m}, \ 2^{\text{cp}(\mathbf{m})})$.

Proof. We first show that $\text{card}(E_M) \leq 2^{\text{cp}(\mathbf{m})}$. By Lemma 1, cdstr is a one-to-one map from the invariant equivalences on $\wp(M)$ into sets of quadruples of cardinals $\leq \mathbf{m}$. But the number of such quadruples is at most $2^{\text{cp}(\mathbf{m})}$; and hence $\text{card}(E_M) \leq 2^{\text{cp}(\mathbf{m})}$.

We next show that $\text{card}(E_M)$, $\text{card}(C_M) \geq 2^{\text{cp}(\mathbf{m})}$. Given a set Γ of cardinals $\leq \mathbf{m}$, define $\equiv_\Gamma$ on $\wp(M)$ by: $C \equiv_\Gamma D$ iff $\text{card}(C)$ and $\text{card}(D)$ both belong to Γ or both fail to belong to Γ. Clearly, for each Γ, $\equiv_\Gamma$ is an invariant equivalence; and, clearly, it is non-inflationary. The relation $\equiv_\Gamma$ divides $\wp(M)$ into two cells, one containing all subsets of M of cardinality in Γ and the other containing all other subsets of M. Thus for each Γ, $X_\Gamma = \{C \subseteq M: \text{card}(C) \in \Gamma\}$ is a strictly admissible cell. But $\text{card}(\{\Gamma: \Gamma$ a set of cardinals $\leq \mathbf{m}\} = 2^{\text{cp}(\mathbf{m})}$; and for distinct Γ, the X_Γ are distinct. So $\text{card}(E_M)$, $\text{card}(C_M) \geq 2^{\text{cp}(\mathbf{m})}$. This establishes (i) (and part of (ii)).

We now show that $\text{card}(C_M) \geq \mathbf{m}$. Define $\approx_s$ on $\wp(M)$ by: $C \approx_s D$ iff either C and D are singleton and the same or C and D are not singleton. Clearly, $\approx_s$ is invariant and non-inflationary; and so each $\{\{x\}\}$ for $x \in M$ is a strictly admissible cell. But there are $\mathbf{m}$ such cells.

We finally show that $\text{card}(C_M) \leq \max(\mathbf{m}, \ 2^{\text{cp}(\mathbf{m})})$. Since each equivalence in E_M induces a partition with at most $\mathbf{m}$ cells, $\text{card}(C_M) \leq \text{card}(E_M).\mathbf{m} = 2^{\text{cp}(\mathbf{m})}.\mathbf{m} = \max(2^{\text{cp}(\mathbf{m})}, \mathbf{m})$. This establishes (ii).

We say that a cardinal $\mathbf{c}$ is *unsurpassable* if $2^{\text{cp}(\mathbf{c})} \leq \mathbf{c}$. The unsurpassable cardinals are the counterpart in our general theory of abstraction to the inaccessible cardinals of ZF. $\aleph_0$ is not unsurpassable since $\text{cp}(\aleph_0) = \aleph_0$. On the other hand, given CH, $2^{\text{cp}(\aleph_1)} = 2^{\aleph_0} = \aleph_1$ and hence $\aleph_1$ will be unsurpassable.

From Lemma 2, we immediately obtain:

Theorem 3. For $\mathbf{m} = \text{card}(M)$ transfinite, the following three conditions are equivalent:
  (i) $\mathbf{m}$ is unsurpassable;
  (ii) $\text{card}(E_M) \leq \mathbf{m}$;
  (iii) $\text{card}(C_M) \leq \mathbf{m}$.

Similar results can be proved using the notion of internal invariance instead of invariance *simpliciter.* Recall that the internal cardinality distribution Icdstr(C, D) of two subsets C and D of M is $<$ card$(D - C)$, card$(C \cap D)$, card$(C - D) >$. We use $I\!E_M$ and IC_M for the corresponding sets of the strictly I-acceptable equivalences and of the strictly I-admissible cells that belong to their partitions. Then clearly the previous upper bounds on card(E_M) and card(C_M) still apply; and the lower bounds also apply given that the equivalences that were used in the constructions were all I-invariant.

Call a §-model M *replete* (*I-replete*) if, for each invariant (resp. I-invariant) non-inflationary equivalence $\approx$ on $\wp(M)$ there is a $§ \in §$ such that $\equiv_§$ is $\approx$. From the theorem we immediately obtain:

Corollary 4. Let M be an infinite domain of urelements and let $\mathbf{m} = \text{card}(M)$. Then the following conditions are equivalent:

   (i) $\mathbf{m}$ is unsurpassable;

  (ii) there is a replete and weakly separated model M with domain M;

 (iii) there is a replete and strictly separated model M with domain M.

In the light of this result it is of interest to ascertain which cardinals are unsurpassable. Note that cp$(\mathbf{c}) = \mathbf{c}$ iff $\mathbf{c} = \aleph_0$ or $\mathbf{c} = \aleph_c$. From this it follows that no cardinal $\mathbf{c}$ for which '$\mathbf{c} = \aleph_0$ or $\mathbf{c} = \aleph_c$' is unsurpassable. But what of the other cardinals? There are two extreme possibilities. One is that they are all unsurpassable; the other is that none are. The first possibility is consistent with ZFC since it follows from GCH; and the second is consistent with ZFC by a result of Foreman and Woodin (1991) according to which it is consistent to assume that $2^\mathbf{c}$ is weakly inaccessible for each transfinite cardinal $\mathbf{c}$ as long as ZFC + the existence of a suitable supercompact cardinal is consistent.

Relativized versions of these results, with elements from a given subset K of M kept fixed, can also be proved. We say that a subset of $\wp(M)$ is a *K-cell* if it is a member of $P_\approx$ for some K-invariant equivalence $\approx$ on M and that a K-cell is *K-admissible* if it is a member of $P_\approx$ for some K-acceptable equivalence $\approx$ over M. The *cardinality distribution* cdstr$_K(C, D)$ *of* C *and* D *relative to* K is taken to be the sextuple $(K_1, K_2, \mathbf{c}, \mathbf{d}, \mathbf{e}, \mathbf{f})$, where $K_1 = C \cap K, K_2 = D \cap K$, and $(\mathbf{c}, \mathbf{d}, \mathbf{e}, \mathbf{f}) = \text{cdstr}(C \cap (M - K), D \cap (M - K))$ (relative to $(M - K)$

as domain). The $\mathrm{cdstr}_K(X)$ for X a K-cell is defined in the same way as before. We then have, in analogy to Lemma 1, that K-invariant equivalences with the same K-cardinality distributions will be the same.

Let $C_{M;K}$ be the set of K-admissible cells over M and let $E_{M;K}$ be the set of non-inflationary K-invariant equivalences on $\wp(M)$. Then in analogy to Lemma 2 and Theorem 3, we have:

Theorem 5. For $\mathbf{m} = \mathrm{card}(M)$ transfinite and K a subset of M of cardinality $\mathbf{k} < \mathbf{m}$:

(i) $\mathrm{card}(E_{M;K}) = \max(2^{\mathrm{cp}(\mathbf{m})}, 2^{2^{\mathbf{k}}})$;

(ii) $\mathrm{card}(C_{M;K}) = \max(\mathbf{m}, 2^{\mathrm{cp}(\mathbf{m})}, 2^{2^{\mathbf{k}}})$.

Proof. Much as for Lemma 2. To show that $\mathrm{card}(E_{M;K})$, $\mathrm{card}(C_{M;K}) \geq 2^{2^{\mathbf{k}}}$, we let $\equiv_X$, for each subset X of $\wp(K)$, be the relation defined on $\wp(K)$ by: $C \in X$ iff $D \in X$.

Corollary 6. Suppose that $\mathbf{m} = \mathrm{card}(M)$ is transfinite and K is a subset of M of cardinality $\mathbf{k} < \mathbf{m}$. Then the following three conditions are equivalent:

(i) $\max(2^{\mathrm{cp}(\mathbf{m})}, 2^{2^{\mathbf{k}}}) \leq \mathbf{m}$;

(ii) $\mathrm{card}(E_{M;K}) \leq \mathbf{m}$;

(iii) $\mathrm{card}(C_{M;K}) \leq \mathbf{m}$.

Proof. Straightforwardly from the theorem.

Where $\mathbf{k}$ is a cardinal $\leq \mathbf{c}$, call a §-model M $\mathbf{k}$-*replete* (*internally* $\mathbf{k}$-*replete*) if, for each K-invariant (resp. internally K-invariant) non-inflationary equivalence $\approx$ on $\wp(M)$, where K is a subset of M of cardinality $< \mathbf{k}$, there is a § $\in$ § for which $\equiv_\S$ is $\approx$. In analogy to Corollary 4, we have:

Corollary 7. For $\mathbf{k}$ a cardinal $\leq \mathbf{m} = \mathrm{card}(M)$, there is a (internally) $\mathbf{k}$-replete and (weakly, strictly) separated model M with domain M iff $2^{\mathrm{cp}(\mathbf{m})} \leq \mathbf{m}$ and $\mathbf{m}^{\mathbf{l}}, 2^{2^{\mathbf{l}}} \leq \mathbf{m}$ for each $\mathbf{l} < \mathbf{k}$.

Note that, given GCH, the condition that $\mathbf{m}^{\mathbf{l}} \leq \mathbf{m}$ will be automatically satisfied when $\mathbf{m}$ is a successor cardinal and the condition that $2^{2^{\mathbf{l}}} \leq \mathbf{m}$ will be equivalent to the condition that $(\mathbf{l}^+)^+ \leq \mathbf{m}$.

We say that an equivalence $\approx$ on $\wp(M)$ is *relatively* $(I\text{-})$ *invariant* if it is (*internally*) K-invariant for some exponentially small subset K of M; and we say that it is *broadly* $(I\text{-})acceptable$ (or $(I\text{-})acceptable$, *simpliciter*) if it is non-inflationary and predominantly (I-)invariant. There is, of course, no most refined broadly acceptable equivalence $\approx$; since, for whichever exponentially small $K \approx$ was K-acceptable, there would be a more refined K^+-acceptable equivalence for any proper superset K^+ of K. A cell is said to be *broadly (I-)admissible* (or *(I-)admissible, simpliciter*) if it is induced by an (I-)acceptable equivalence. Thus when $\mathbf{k} = \text{exbd}(\mathbf{m})$, each admissible cell will be represented by an abstract in the $\mathbf{k}$-replete model M.

The I-admissible cells have the following interesting property:

Theorem 8. A cell on the infinite domain M is I-admissible iff it is relatively I-invariant, i.e. stays the same under all internally K-invariant permutations on M for some exponentially small subset K of M.

Proof. Suppose first that the cell X is internally K-admissible for K an exponentially small subset of M. Thus for some internally K-acceptable equivalence $\approx$ over M, X is a member of the partition induced by $\approx$. For any subset P of K, let $X_P = \{C \subseteq (M-K): C \cup P \in X\}$ and let $\approx_P$ be the relation on $M - K$ defined by: $C \approx_P D$ if $C \cup P \approx D \cup P$. For any subsets P and Q of K, define $\leftrightarrow_{P, Q}$ on subsets C and D of $M-K$ by: $C \leftrightarrow_{P, Q} D$ if $C \cup P \approx D \cup Q$. It is readily verified that $\leftrightarrow_{P, Q}$ is a biequivalence and that the relation $\leftrightarrow_{L;P, Q}$ of connection on the left is coincident with $\approx_P$ and that the relation $\leftrightarrow_{R;P, Q}$ of connection on the right is coincident with $\approx_Q$.

We establish the following facts (for P and Q subsets of K):

(1) If $C \approx_P D$, for C very different from D, then $\approx_P$ is numeric on card(C).

Proof. It is clear that the equivalence $\approx_P$ is an I-invariant equivalence over $M - K$. The result therefore follows from Corollary 7.13.

(2) If $C \leftrightarrow_{P, Q} D$, for C very different from D, then $\approx_P$ is numeric on card(C).

Proof. From Theorem 7.15.

(3) If X_P contains an exponentially large set C then it contains a set D very different from C.

Proof. $\approx_P$ is a strictly I-acceptable equivalence over $M - K$. So by Corollary 7.7, it is numeric on card(C). But then we may readily find a D that is equinumerous with C and yet very different from C.

(4) If $X^- = \cup\{X_P : P \subseteq K\}$ contains an exponentially large set or two sets that are very different, then each X_P is numeric, i.e. is closed under equinumerosity.

Proof. If X^- contains an exponentially large set C then it follows from (3) that it contains two sets that are very different; and so we need only consider the one case in which X^- contains two sets C and D that are very different. Take now any set E in X_P for $P \subseteq K$. Then E is either very different from C or from D. Without loss of generality, assume the former. If $D \in X_P$, then the desired result follows from (1). If $D \in X_Q$, for Q distinct from P, then the desired result follows from (2).

We are now in a position to establish the direction from left to right. By (4), we need only consider the following two cases:

(i) Each X_P is numeric. Let $X^P = \{C \in X : C \cap K = P\}$. Then each X^P is internally K-invariant; and so $X = \cup\{X^P : P \subseteq K\}$ is K-invariant.

(ii) X^- contains only exponentially small sets and any two are almost the same. We may show that the members of any of the two non-empty sets X_P are the same. For suppose $C \in X_P$ and $D \in X_Q$, with $C \neq D$. Then $C \leftrightarrow_{P,Q} D$. Let $\mathbf{d} = \text{card}(C \sim D)$. Then by Theorem 7.15, $C \approx_P E$ for any E for which $E \equiv_{\mathbf{d}+} C$; and so, since $D \equiv_{\mathbf{d}+} C, D \in X_P$.

We now distinguish two subcases. (a) X^- is the singleton $\{C\}$. But then X will be $K \cup C$ invariant, with $K \cup C$ exponentially small. (b) X^- is not singleton. Given that X^- is a cell induced by any of the the $\approx_P$ for which X_P is non-empty, it follows from Theorem 7.12 that X^- will be an equivalence class $|C|$ wrt to one of the relations $\equiv_{\mathbf{d}}$. But then, again, X will be $K \cup C$ invariant, with $K \cup C$ exponentially small.

For the other direction, which is much more straightforward, suppose that X is an internally K-invariant cell for some exponentially small subset K of M. Define $\approx$ on $\wp(M)$ by: $C \approx D$ iff C and D both belong or both fail to belong to X. Then it is clear that $\approx$ is non-inflationary and internally K-invariant.

An analogous result does not hold for strict admissibility. For when C is exponentially small, $\{C\}$ will in general be strictly I-admissible and yet not I-invariant.

9. Internalized Proofs

For the purposes of the present section, we take ourselves to be working within a particular abstraction theory T^ϕ, with ϕ §-free.

We will suppose that the theory contains a well-ordering principle, WO, which states that there is a well-ordering of the whole universe, and an axiom of infinity, Inf, which states that there exists an ordering without any last element.

We show how the main negative results of sect. 7 can be formalized within the resulting theory. This means that the results will have a claim on any abstraction theorist who accepts the underlying second-order logic and that they do not rest upon her adopting a 'full' Platonic stand on concepts.

Theorem 1. In any logical abstraction theory $T^\phi + WO + Inf$, Unstb(C) & Unstb(D) & (C beq D) $\rightarrow$ §C = §D is a theorem.

Proof. Let me briefly sketch how the proof of Theorem 7.6 may be formalized within $T^\phi + WO + Inf$. We need to establish appropriate internal analogues of Lemmas 7.3 and 7.5. For simplicity, let us concentrate on the case of bifurcators. We then need to show:

(1) If two bifurcatory concepts C and D are identified by ϕ, with C very different from both D and its complement, then all bifurcators are identified; and

(2) If no two bifurcatory concepts C and D, with C very different from both D and its complement, are ever identified, then there is definable one-to-one correspondence taking the bifurcatory concepts into objects.

The condition that the bifurcatory concepts C and D are very different can be expressed as C – D or D – C being equinumerous with C. We cannot say in the second-order language L^ϕ that there exists a one-to-one correspondence between concepts and objects; but we will be able to produce, on the basis of a proof, a formula $\chi(C, x)$ (containing parameters) with the properties of a one-to-one correspondence.

To formalize the proofs of (1) and (2), we may use Lemma 3.7 to obtain the 'internalized' counterpart to the claim that the relation defined by ϕ is invariant. We then need to imitate the reasoning about cardinals and the various constructions on sets, some of which takes place outside the underlying domain M. This will require the Schroeder-Bernstein Theorem in the form: if there is a one-to-one correspondence from C into D and a one-to-one correspondence of D into C, then there is a one-to-one correspondence of C onto D. The theorem in this form can be proved along the lines of the proof in Drake (1974: 49). For as we have seen, the fixed-point theorem can be proved within L^2 (Theorem 3.2); and it is then an easy matter to define the relevant 'functions' on concepts and to establish the relevant properties. (Cf. Shapiro 1991: 102–3).

For some of the proofs, we will need Well-Ordering. Given this assumption, the ordinals can be identified with the objects under the posited well-ordering and the cardinals can be identified with the surrogates for the initial orderings. It can be then be shown that the objects falling under any concept can be put into one-to-one correspondence with an initial segment of the 'ordinals' and hence assigned a 'cardinal'.

To divide the objects falling under an infinite concept into two equal-sized parts, we take advantage of the one-to-one correspondence onto an initial segment of the 'ordinals'. A definition by transfinite induction can then be used to divide the ordinals into two equal-sized parts; and this then induces a corresponding division of the objects.

The original proof of (2) requires that we take **m** copies of the universe M. Clearly, this cannot be done in L^2. But what we can do instead is to set up a one-to-one correspondence from 'pairs' of objects from the domain into the whole domain. Given Well-Ordering, there is no difficulty in defining such a correspondence. A 'copy' of M is then obtained by taking the objects that correspond to all the pairs (x_0, y) for a fixed x_0.

Recall that $\phi_0(C, D)$ is the object-language version of the basal equivalence $\approx_0$. An equivalent form of the result is:

Corollary 2. In any theory $T^\phi + WO + Inf$, with logical ϕ, the formula $\phi_0(C, D) \rightarrow \phi(C, D)$ is a theorem.

Let $Bf(C)$ be the formula in L^2 which says that there is a one-to-one correspondence between the objects that fall under C and those that fail to fall under C. Since $Bf(C) \rightarrow Unstb(C)$ and $Bf(C) \& Bf(D) \rightarrow C$ beq D are theorems of L^2, as a special case of the theorem we obtain:

Corollary 3. In any logical abstraction theory $T^\phi + WO + Inf$, the formula $Bf(C) \& Bf(D) \rightarrow \S B = \S D$ is a theorem.

These results have consequences for restricted theories of extensional abstraction. A natural response to the paradoxes is to restrict the concepts that are capable of having an extension. Recall that, when $\psi = \psi(C)$ is a formula containing C as its sole free variable, $T^{\psi, \equiv}$ is the result of adding to L^2 the axiom:

Ext$_\psi$ · ψ(C) & ψ(D) → (§C = §D ↔ C ≡ D).

We may now prove within the theory of restricted abstraction that no unsuitable concepts can be subject to extensional abstraction:

Corollary 4. In any theory $T^{\psi, \equiv}$ + WO + Inf, with logical ψ, ψ(C) → Stb(C) (and hence ψ(C) → ¬Bf(C)) is a theorem.

Proof. Let ϕ = ϕ(C, D) be the formula: (ψ(C) & ψ(D) & C ≡ D) ∨(¬ψ(C) & ¬ψ(D)). By Corollary 2, we can prove the formula Unstbt (C) & Unstbt(D) & C beq D → ϕ(C, D) within T^ϕ +WO + Inf. But the § of L^ϕ can be 'defined' within $L^{\psi, \equiv}$ by 'identifying' §C, for the C that do not conform to ψ, with an arbitrary object from the domain (which can always be 'detached' from the rest of the universe by means of WO and Inf). Theorems of T^ϕ then translate into theorems of $T^{\psi, \equiv}$; and hence Unstb(C) & Unstb(D) & C beq D → ϕ(C, D) can be proved within $T^{\psi, \equiv}$ + WO + Inf. We can then prove ψ(C) → Stb(C) within $T^{\psi, \equiv}$ by transcribing the following informal proof:

Suppose there were an unsuitable C that conformed to ψ. Choose a D that is biequinumerous with but not coextensive with C. Now D is unsuitable; and by the invariance of ψ, it also conforms to ψ. But then from Unstb(C) & Unstb(D) & C beq D → ϕ(C, D), it follows that C and D are coextensive. A contradiction.

The only property of ψ required for the above proof is that it be invariant. Hence if we add a third-order monadic predicate constant S to the language $L^§$, S(C) → Stb(C) can be proved in the theory with the axioms: WO, Inf, S(C) & S(D) → (§C = §D ↔ C ≡ D), and (S(C) & C beq D) → S(D) as axioms. Thus whatever logical restriction be imposed on the concepts subject to extensional abstraction, it can be *proved* within the resulting theory that they must all be suitable; either their extensions or their counter-extensions must be exponentially small.

The above results can be used to establish the inconsistency of a large number of logical abstraction principles; for all that need be shown is that the principles in question fail to identify unsuitable concepts that are biequinumerous. They can therefore be regarded as providing a highly general form of the paradoxical reasoning, though subject to the limitation that the criterion of identity be logical and that Well-Ordering and Infinity be assumed. It would be desirable to produce versions of these results that did not depend upon these assumptions.

IV

The General Theory of Abstraction

WE develop a general theory of abstraction, one that is intended to account for the existence and behaviour of abstracts in general, and not of any kind of abstract in particular. The material is in three sections: first we outline the systems; then we look at their models; and finally we show how they serve to provide a foundation for both arithmetic and analysis.

1. The Systems

Since we want to talk about all equivalence relations on concepts, the most natural framework for our theory is third-order rather than second-order logic. Most of our previous definitions for second-order logic will extend in the obvious way to third-order logic; but to avoid confusion, we will use boldface to distinguish the third-order terms from the others. Later it will be shown how a version of the theory can be given within the framework of second-order logic.

The sole extra-logical primitive of our theory is the three-place predicate Abstr, which applies to terms for an object, a concept, and a relation, in that order. The predications formed from Abstr may be written in the form t $\text{Abstr}_{\mathbf{R}}$ C; and the intended reading is that t is an abstract of C with respect to $\mathbf{R}$. Note that in contrast to a functional notation, such as $\S_{\mathbf{R}}(C)$, the predications formed from Abstr can be used without any commitment to the existence of abstracts.

The following definitions will be useful (some have already been given but are here restated in an appropriately modified form):

Unrelativized Abstraction: x Abstr C for $\exists \mathbf{R}(\text{x Abstr}_{\mathbf{R}} \text{ C})$;
$\mathbf{R}$-Abstractable concepts: $\text{Abstr}_{\mathbf{R}}(C)$ for $\exists x(\text{x Abstr}_{\mathbf{R}} \text{ C})$;
$\mathbf{R}$-Abstracts: $\text{Abstr}_{\mathbf{R}}(x)$ for $\exists C(\text{x Abstr}_{\mathbf{R}} \text{ C})$;
Abstract: Ab(x) for $\exists \mathbf{R} \exists C(\text{x Abstr}_{\mathbf{R}} \text{ C})$;
Applicability: App($\mathbf{R}$) for $\forall C(\text{Abstr}_{\mathbf{R}}(C))$;

Invariance: Invar(**R**) for $\forall$P, C, D, C′, D′ [Perm(P) & **R**(C, D) & $\forall$x, x′ (Pxx′ → (Cx ↔ C′x′) & (Dx ↔ D′x′)) → **R**(C′, D′)];

Internal Invariance: I-invar(**R**) for $\forall$P, C, D, C′, D′[1–1(P) & $\forall$x ((x ∈ Dm(P) ↔ Cx ∨ Dx) & **R**(C, D) & $\forall$x, x′(Pxx′ → (Cx ↔ C′x′) & (Dx ↔ D′x′)) → **R**(C′, D′)];

K-invariance: Inv$_K$(**R**) for $\forall$P, C, D, C′, D′ [Perm(P) & $\forall$x (Kx → Pxx) & **R**(C, D) & $\forall$x, x′(Pxx′ → (Cx ↔ C′x′) & (Dx ↔ D′x′)) → **R**(C′, D′)].

Internal K-invariance: I-invar$_K$(**R**) for $\forall$P, C, D, C′, D′ [1–1(P) & $\forall x$((x ∈ Dm(P) ↔ Cx ∨ Dx ∨ Kx) & $\forall$x (Kx → Pxx) & **R**(C, D) & $\forall$x, x′(Pxx′ → (Cx ↔ C′x′) & (Dx ↔ D′x′)) → **R**(C′, D′)];

Power: D powers C for $\exists$**R**$\forall$C′(C′ ⊆ C → $\exists$y(Dy & $\forall$x (Rxy ↔ C′x)));

Modesty: Modest(C) for $\exists$**D**$\exists$E(D powers C & E powers D).

Predominant invariance: Prdinv(**R**) for $\exists$K(Expsmall(K) & Modest(K) & *Inv$_K$*(**R**));

Predominant internal invariance: I-Prdinv(**R**) for $\exists$K (Expsmall(K) & Most(K) & I-inv$_K$(**R**));

Non-inflationary: Noninfl(**R**) for $\exists$P[$\forall$C$\exists$xP(C, x) & $\forall$C, D, x (**P**(C, x) & **P**(D, x) → **R**(C, D)].

Related notions may be defined in an analogous way; and their formal definitions will not always be given.

A concept is abstractable when it yields an abstract; a third-order relation is invariant when its extension remains the same under any permutation; and a third-order relation (which we conceive as a method for identifying concepts) is non-inflationary when the non-identified concepts can be mapped one-to-one into objects.

Let me first give a formulation of the basic theory. I shall then consider various ways in which it may be modified. There are three axioms in all. The first gives necessary and sufficient conditions for two abstracts to be the same:

Identity: (xAbstr$_R$ C & y Abstr$_S$ D) → [x = y ↔ (**R**(C, D) & $\forall$E(**R**(C, E) ↔ S(C, E)))].

Thus this axiom says that two abstracts, possibly associated with different methods of abstraction **R** and **S** and different concepts C and D, will be the same just in case each method identifies the two concepts and each method identifies the same concepts with one of the given concepts (given the second conjunct there is no need to state S (C,D) since it follows from the first conjunct).

The second axiom gives a sufficient condition for abstractions to exist:

Existence: Eq($\mathbf{R}$) & Inv($\mathbf{R}$) & Noninfl($\mathbf{R}$) $\rightarrow$ App($\mathbf{R}$).

Thus according to this axiom, any invariant and non-inflationary equivalence is an applicable form of abstraction; it will yield abstracts for each first-order concept.

None of the conditions in the antecedent can sensibly be dropped. This is obvious for the first. If we drop the third, the requirement that $\mathbf{R}$ be non-inflationary, then a contradiction can be proved in the manner of Russell's paradox by letting $\mathbf{R}$ be the relation of coextensiveness. If we drop the second conjunct, the requirement that $\mathbf{R}$ be invariant, then a contradiction can be proved by a hyperversion of the Russellian reasoning. For with each concept C we may associate a non-inflationary equivalence $\equiv_C$ according to the condition that $D \equiv_C E$ iff $D \equiv C \leftrightarrow E \equiv C$. (This equivalence divides the universe of concepts into those that are coextensive with C and those that are not.) With respect to each such equivalence $\equiv_C$, there will exist an abstract ex(C) of C. We may define a concept C_r by the condition that $C_r x$ iff x is an object of the form ex(C) which does not fall under C. Considering the question of whether or not ex(C_r) falls under C_r then allows us to derive a contradiction.

The third axiom states that either no concepts are abstractable (with respect to a given method of abstraction) or all are:

Application: Abstrs$_\mathbf{R}$(C) $\rightarrow$ Abstrs$_\mathbf{R}$(D).

The various axioms can be weakened or strengthened in various respects. The identity axiom implies a necessary and sufficient condition for an abstract x of one kind $\mathbf{R}$ to be the same as an abstract y of another kind $\mathbf{S}$; and this is that each method of abstraction associate the same concepts with the two objects. It is important for the deductive development of the theory that the condition be accepted as necessary; if the associated concepts are distinct then the abstracts themselves must be distinct. The requirement of sufficiency, however, can be dropped; and the question of what might plausibly take its place has been considered in sect. I.5.

The existence axiom only provides for the existence of invariant methods of abstraction. It would be desirable to extend it to various non-invariant methods even though, on pain of contradiction, we cannot admit them all. Our earlier results on K-invariance suggest a natural way in which this might be done:

Broad Existence: $Eq(\mathbf{R})$ & $Prdinv(\mathbf{R})$ & $Noninfl(\mathbf{R}) \rightarrow App(\mathbf{R})$.

Under a generative conception of abstraction, we would want to replace the existence axiom by the weaker principle:

Internal Existence: $Eq(\mathbf{R})$ & $I\text{-}inv(\mathbf{R})$ & $Noninfl(\mathbf{R}) \rightarrow App(\mathbf{R})$.

Similarly, the axiom of broad existence might be replaced with:

Broad Internal Existence: $Eq(\mathbf{R})$ & $I\text{-}Prdinv(\mathbf{R})$ & $Noninfl(\mathbf{R}) \rightarrow App(\mathbf{R})$.

We might also insist that all methods of abstraction are to be predominantly invariant. Assuming internality, the required axiom would then take the form:

Exclusivity: $App(\mathbf{R}) \rightarrow I\text{-}Prdinv(\mathbf{R})$.

Using this axiom, the arrow in the preceding existence axiom can then be reversed.

A principle of minimality might also be adopted. In the light of the discussion at the end of sect. III.5, this could be stated in the form:

Minimality: $\neg \exists D \{ \exists x \neg Dx$ & $\forall x \, (I(x) \rightarrow Dx)$ & $\forall C \subseteq D, x, \mathbf{R}$ $[x \; Abstr_{\mathbf{R}}C \rightarrow \exists y[Dy$ & $\forall C \subseteq D \; (y \; Abstr \, C \leftrightarrow x \; Abstr_{\mathbf{R}}C)]\}$.

In the present context, we may however, state the axiom in a stronger form. Let us note, in this regard, that the definitions of being exponentially small and the like can be relativized to a subdomain, as given by a concept D. We use $Expsmall^D(C)$ and the like for these relativized notions. We define the relativized notions of K-acceptability, acceptability, and strict acceptability by:

$K - Acc^D(\mathbf{R})$ for $[Eq(\mathbf{R})$ & $\text{Non-inflationary}^D(\mathbf{R})$ & $(K \subseteq D)$ & $\forall B_1, B_2, C_1, C_2, P(1\text{--}1(P)$ & $\forall x \, (Kx \rightarrow Pxx)$ & $\mathbf{R}(B_1, B_2)$ & $B_1 \rightarrow_P C_1$ & $B_2 \rightarrow_P C_2 \rightarrow \mathbf{R}(C_1, C_2)]$;

$Acc^D(\mathbf{R})$ for $\exists K(Expsmall^D(K)$ & $Modest^D \, (K)$ & $K - Acc^D(\mathbf{R}))$;

$S - Acc^D(\mathbf{R})$ for $\exists K(\forall x \neg Kx$ & $K - Acc^D(\mathbf{R}))$.

Then the alternative formulation is:

Strong Minimality: $\neg\exists D\{\exists x\neg Dx$ & $\forall x(I(x)\to Dx)$ & $\exists x\exists y\ (x\neq y)$ & $\forall C\subseteq D, x, \mathbf{R}(Acc^D(\mathbf{R})$ & $x\ Abstr_\mathbf{R}C\to\exists y[Dy$ & $\forall C\subseteq D$ $(y\ Abstr\ C\ \leftrightarrow\ x\ Abstr_\mathbf{R}C)]$.

Thus closure is here required only with respect to the means of abstraction that are acceptable within the given subdomain. In case only logical definitions of the abstracts are allowed, the axiom should take the same form but with $Acc^D(\mathbf{R})$ replaced by $S-Acc^D(\mathbf{R})$.

The Application axiom is rather strong and might be weakened by allowing restricted methods of abstraction, i.e. ones that are only defined on the concepts in the field of their relation. The principle should then take the form:

Weak Application: $(Abstr_\mathbf{R}(C)$ & $C\in Fld(\mathbf{R}))\to\forall D(D\in Fld(\mathbf{R})\to Abstr_\mathbf{R}(D))$.

The existence axiom must also then be modified so as to allow for the existence of restricted methods of abstraction. However, there is no great disadvantage in working with the tighter principle. For any restricted or partially defined method of abstraction can be identified with an unrestricted method that identifies all the unabstractable concepts with one of the abstractable concepts.

The original application axiom or its variants are not really essential for the intended deductive development of the system, although they do make it simpler.

We turn to the second-order formulation of the theory. This uses schemes in place of universal principles. The primitive non-logical predications of the theory now take the form t $Abstr_{[C,\,D]\ \phi}$ E, where t is an object term, C and D are distinct concept variables, ϕ is a formula, and E is a concept term. In the resulting formula, the variables C and D are taken to be bound.

The intuitive meaning of the formula t $Abstr_{[C,\,D]\phi}$ E is given by t $Abstr_\mathbf{R}$ E, where $\mathbf{R}$ is the relation $(\lambda C, D)\phi$. However, in the formula t $Abstr_{[C,\,D]\phi}$ E, we treat the complex $[C,D]\phi$ as a part of the whole notation and not as a meaningful constituent in its own right. Thus the only second-order relations we can talk about are those that can be specified by a formula (with or without parameters).

The axioms are the same as before but with $\mathbf{R}$ now taken to be a complex of the form $[C,D]\phi$. Thus whereas the axioms, under the previous formulation, could be taken to be single sentences universally quantified by a third-order variable **R**, they must now be taken to be schemes.

The retrenchment from the third- to the second-order formulation involves some complications. First, the application **R**[E,F] of **R** to the concept terms E and F must now be understood as the simultaneous substitution of E and F for C and D in φ. Second, the original definition of Noninfl(**R**) cannot be used since it is third order. Instead of existentially quantifying over the third-order correlation **P** between concepts and objects, we may suppose that **P** is specified by a formula (possibly with parameters). Third, the Minimality Axiom, which involves an inelimable reference to third-order relations, must be dropped. It is not clear which second-order axioms, if any, should be put in its place.

Let us denote the basic third-order theory, consisting of the axioms Identity, Broad Internal Existence, and Application, by GA; and let us denote the result of adding Exclusivity and Strong Minimality to GA by GA$^+$. We use GA^2 for the second-order counterpart of GA; and we use SGA and SGA^2 for the systems that result from replacing the broad existence axiom in GA and GA^2, respectively, by its normal invariant form. For many purposes, the differences between the systems will not matter; and, in particular, the derivation of arithmetic and analysis given in sect. 3 can be carried out equally well in GA$^+$ and in SGA^2.

It follows from theorem III. 8.7 that GA will have a standard model of cardinality **c** just in case **c** is an unsurpassable cardinal. Since $\aleph_0$ is surpassable, it will not have a standard model of cardinality $\aleph_0$. But given CH, it will have a standard model of cardinality $\aleph_1$; and so from the consistency of CH will follow the consistency of GA. In the next section, we shall construct some natural models for GA on any domain of an unsurpassable cardinality and thereby show that the various additional axioms can also be satisfied.

We state the following elementary results:

Lemma 1. The following are theorems of GA (and GA^2):
(i) x Abstr$_R$C & y Abstr$_R$D → (x = y ↔ **R**(C, D));
(ii) ∃C(Abstrs$_R$(C)) → Eq(**R**);
(iii) x Abstr$_R$C → (**R**(C, D) ↔ x Abstr$_R$D);
(iv) Abstr$_R$(x) & Abstr$_S$(y) → [x = y ↔ ∀E(x Abstr$_R$E ↔ y *Abstr$_S$E*);
Proof. The proofs are stated informally but are readily formalized.

(i) Let **R** and **S** in Identity be the same. The second conjunct on the right is then trivially satisfied.

(ii) Given $\exists C\ (\text{Abstr}_{\mathbf{R}}(C))$, it follows by Application that $\text{App}(\mathbf{R})$. Now to prove $\text{Sym}(\mathbf{R})$, suppose $\mathbf{R}(C, D)$. By $\text{App}(\mathbf{R})$, x $\text{Abstr}_{\mathbf{R}}$ C and y $\text{Abstr}_{\mathbf{R}}$ D for some x and y. By (i) and the fact that $\mathbf{R}(C,D)$, $x = y$. But then by another application of (i), $\mathbf{R}(D, C)$. $\text{Trans}(\mathbf{R})$ and $\text{Refl}(\mathbf{R})$ are proved similarly. (Note that the assumption of $\text{App}(\mathbf{R})$ is essential to the proof; for there is no need for inapplicable relations to be equivalence relations.)

(iii) For the left to right direction, assume x $\text{Abstr}_{\mathbf{R}}$ C and $\mathbf{R}(C, D)$. It then follows by Application that y $\text{Abstr}_{\mathbf{R}}$ D for some y. But $x = y$ by (i); and hence x $\text{Abstr}_{\mathbf{R}}$ D.

For the right to left direction, use (i).

(iv) We may write the antecedent in the form x $\text{Abstr}_{\mathbf{R}}C$ & y $\text{Abstr}_{\mathbf{S}}$ D. Assume its truth in this form. Then by (iii), the final consequent of (iv) is equivalent to $\forall E(\mathbf{R}(C,E) \leftrightarrow \mathbf{S}(D, E))$. Hence given Identity, it suffices to show the equivalence of $\forall E(\mathbf{R}(C, E) \leftrightarrow \mathbf{S}(D, E))$ to $\mathbf{R}(C, D)$ & $\forall E(\mathbf{R}(C, E) \leftrightarrow \mathbf{S}(C, E)))$. Assume the first formula. Now $\mathbf{S}(D, D)$ by (ii); and so $\mathbf{R}(C, D)$. Also $\mathbf{R}(C, C)$, by (ii) again; and so $\mathbf{S}(D,C)$. But then by transitivity, $\mathbf{S}(C,E) \leftrightarrow \mathbf{S}(D, E)$; and so $\forall E(\mathbf{R}(C, E) \leftrightarrow \mathbf{S}(C, E))$. Now assume the second formula. We obtain $\mathbf{S}(C, D)$ by instantiation and $\mathbf{S}(D, C)$, from this, by Symmetry. But then $\mathbf{S}(D, E) \leftrightarrow \mathbf{S}(C, E)$ and hence $\forall E(\mathbf{R}(C, E) \leftrightarrow \mathbf{S}(D, E))$.

The principle under (i) provides a criterion for intra-sortal identity while the principle under (iv) provides a criterion for cross-sortal identity, stated directly in terms of abstraction and not the underlying relation. These two principles are combined together in our original Identity Axiom.

Suppose that we were to use $\S_{\mathbf{R}}C$ to designate the abstract of C with respect to the equivalence relation **R**. Then the abstraction principle for **R**-abstracts could be stated in the categorical form:

$$\S_{\mathbf{R}}C = \S_{\mathbf{R}}D \leftrightarrow \mathbf{R}(C, D).$$

From this principle, within a standard logical system, the existence of the **R**-abstract $\S_{\mathbf{R}}C$ of each concept C would then follow. The principle under (i) is a kind of conditional version of the categorical principle in its categorical form. It states that *if* there are **R**-abstracts of C and D, then they will behave in accord with that principle. The existential implications of the categorical principle have led many authors to doubt its analyticity. But the principle in its conditional form is free of any such implications and hence has a much better claim to be considered analytic.

It is of interest to consider the theory that results from combining the general theory of abstraction with ZF. This is the theory that might be adopted by someone who believed both in sets (as given by the cumulative hieararchy) and in abstracts. Let us assume a standard formulation of third-order ZFI (with Choice). There will be a special predicate I for the individuals, but it will not be assumed that the individuals constitute a set. On this is then imposed the third-order theory GA (or a variant thereof). Let us call the result ZFA.

Given the existence of an inaccessible cardinal c_0 and of a greater unsurpassable cardinal c_1, it may be shown that the theory ZFA has a 'standard' model. For we may let the domain I of individuals be of cardinality c_1; and over this domain we may construct a cumulative hierarchy U_α of sets, for $\alpha < c_0$, by requiring the sets generated at each stage to be of cardinality less than c_0. Over the resulting domain U of individuals and sets, we may then construct a standard model of GA.

In this model, the size of I is greater than that of U. It can actually be proved within ZFA that this must hold, i.e. that the abstracts must outnumber the pure sets. This result is of some philosophical significance since it shows that it is impossible for our proponent of sets and abstracts to identify the abstracts, in any reasonable manner, with the sets.

Let us use $Pset(x)$ to mean that x is a pure set (one whose transitive closure contains no urelements). We then have:

Theorem 2. Within the theory ZFA, the formula:

$$\neg\, (\exists R)(1\text{–}1(R)\ \&\ \forall x(Ab(x) \rightarrow x \in Dm(R))\ \&\ \forall y(y \in Rg(R) \rightarrow PSet(y)))$$

is a theorem.

Proof. We sketch an informal version of the proof. Given a non-empty concept C of (ZF) cardinals we may form the sequence s_C of its members in increasing order of magnitude. (s_C in general will, of course, not be a set but a concept or class. It could, for example, be represented by a one-to-one relation from ordinals to cardinals.) We may then order such concepts by the rule:

$C \leq D$ iff either (i) C and D are coextensive, or (ii) s_C is of smaller length than s_D, or (iii) they are of the same length and, at the first point at which they differ, the value of s_C is smaller than the value of S_D.

It is readily shown that $\leq$ is a transitive and connected relation and that the corresponding strict relation $<$ is well-founded. (Note that to express that $<$ is well-founded we need to go to the third order.)

With each non-empty concept of cardinals C we may associate the equivalence relation $\equiv_C$ on concepts, where:

$D \equiv_C E$ iff D and E both constitute sets of a cardinal that falls under C or both fail to constitute sets of a cardinal that falls under C.

With respect to each such equivalence relation $\equiv_C$, we may associate the abstract x_C of a concept D whose extension is of a cardinal that falls under C. (The abstract is the same whichever concept D we choose). When $C < D$, the concepts C and D will be distinct and hence so will the associated abstracts x_C and x_D.

Suppose, for reductio, that there is a one-to-one map from abstracts to pure sets. Then, in particular, there is a one-to-one map from the abstracts x_C to the pure sets (call these sets the *surrogates*). The well-founded ordering $<$ on the concepts C therefore induces a well-founded ordering $<_S$ on the surrogates.

Now this ordering $<_S$ can be used to construct a one-to-one map from surrogates to ordinals. For we may reorder the surrogates according to the rule:

$x < y$ iff either rank $(x) <$ rank(y) or rank$(x) =$ rank(y) and x $<_S$ y.

Each surrogate y will then be preceded by a *set* of surrogates according to this ordering; and so the resulting ordering will induce a one-to-one map from surrogates to ordinals.

We may therefore obtain a corresponding one-to-one map from surrogates to cardinals. But since surrogates are associated with concepts of cardinals, we have a one-to-one map from concepts of cardinals to cardinals; and the reasoning of Russell's paradox may be applied to obtain a contradiction.

There is another, rather different, way in which the theory GA might be related to set theory; for, as mentioned in sect. I.4, the abstracts themselves might be regarded as classes (or extensions) of concepts. Thus given an abstract x, we may take $C \in x$ to hold just in case $\exists R(x \text{ Abstr}_R C)$. Given the Identity Axiom, we may then establish:

Extensionality: $\forall x, y (\forall C(C \in x \leftrightarrow C \in y) \rightarrow x = y)$;

and by formalizing the proof of the easy direction of Theorem III. 8.8 we may also establish:

Comprehension: I-Prdinv(E) $\rightarrow \exists x \forall C(C \in x \leftrightarrow E(C))$,

where predominant invariance for a second-level concept E is defined in the obvious way.

Similarly, within the strict theory SGA, we may show:

Strict Comprehension: I-inv(E) $\rightarrow \exists x \forall C(C \in x \leftrightarrow E(C))$;

and within the corresponding schematic formulations of GA and SGA, we may show:

Schematic Comprehension: I-Prdinv($\psi(C)$) $\rightarrow \exists x \forall C(C \in x \leftrightarrow \psi(C))$;
Strict Schematic Comprehension: I-inv($\psi(C)$) $\rightarrow \exists x \forall C(C \in x \leftrightarrow \psi(C))$.

On the other hand, the predicate $\in$ could itself be taken as a primitive and Extensionality, and along with one of the forms of Comprehension, be adopted as axioms. Let us use STE^2 (theory of extensions) for the weak second-order theory consisting of *Extensionality* and *Strict Schematic Comprehension*; and let us use TE for the strong third-order theory consisting of *Extensionality* and *Comprehension* in its broad third-order formulation.

It is natural to extend the resulting theories to allow for classes of objects as well of concepts. Comprehension would then take the form:

I-Prdinv($\psi(z)$) $\rightarrow \exists x \forall z(z \in x \leftrightarrow \psi(z))$.

However, such an extension would not significantly increase the power of the theory, since objects could always be identified with their singleton concepts and the corresponding form of comprehension on sets of concepts be used in place of a given instance of comprehension on sets of objects.

Within the strong theory TE, one might define what it is for a relation **R** on concepts to be broadly I-acceptable. By formalizing the proof of the hard direction of Theorem III. 8.8 (using the appropriate form of the axiom of choice), one could then prove within the theory itself:

$(\forall \mathbf{R})(\mathbf{R}$ is broadly I-acceptable $\rightarrow (\forall C)(\exists y)\forall D(D \in y \leftrightarrow \mathbf{R}(C, D))$;

and similarly for the schematic versions of the theory.

By adding an appropropriate form of the axiom of choice, it should also be possible to establish the 'equivalence' (or, more exactly, the mutual interpretability) between different versions of the theory of classes and corresponding versions of the theory of abstraction. But this is not a matter that we shall investigate.

2. Semantics

We give what might plausibly be regarded as the standard models for our general theory of abstraction. We concentrate on the case in which individuals are allowed in the definition of the methods of abstraction, although we shall also consider the case in which individuals are not allowed. In either case, the models are constructed according to the generative standpoint described in sect. I.2. Thus starting off with a given class I of individuals, each construction yields a cumulative sequence of domains (denoted respectively by $U_{I,\alpha}$ and $V_{I,\alpha}$); and the intended model (either U_I or V_I) is given by a fixed point of the construction.

We collect together some results on cardinals, which will later be of use. On unsurpassables, we have:

Lemma 1. (i) If $\mathbf{u}$ is unsurpassable, then so is $2^{\mathrm{cp}(\mathbf{u})}$;

(ii) If $\mathbf{u}$ is the smallest unsurpassable cardinal greater than a surpassable cardinal $\mathbf{c}$, then $2^{\mathrm{cp}(\mathbf{u})} = \mathbf{u}$.

Proof. For the purposes of the proof, let us use $\mathbf{u}^\star$ for $2^{\mathrm{cp}(\mathbf{u})}$.

(i) Given that $\mathbf{u}$ is unsurpassable, $\mathbf{u}^\star \leq \mathbf{u}$; so $\mathrm{cp}(\mathbf{u}^\star) \leq \mathrm{cp}(\mathbf{u})$; and so $2^{\mathrm{cp}(\mathbf{u}^\star)} \leq 2^{\mathrm{cp}(\mathbf{u})} = \mathbf{u}^\star$.

(ii) If $\mathbf{u}^\star < \mathbf{c}$, then $2^{\mathrm{cp}(\mathbf{c})} \leq 2^{\mathrm{cp}(\mathbf{u})} = \mathbf{u}^\star < \mathbf{c}$; and so $\mathbf{c}$ is unsurpassable, contrary to supposition. So $\mathbf{u}^\star \geq \mathbf{c}$. But then $\mathbf{u}^\star = \mathbf{u}$, since otherwise $\mathbf{u}^\star$, by (i), will be a smaller unsurpassable than $\mathbf{u}$ that is greater than $\mathbf{c}$.

We call an unsurpassable $\mathbf{u}$ *steady* if $\mathbf{u}^\star = \mathbf{u}$. Let us note that, under the assumption of GCH, the steady unsurpassables are exactly the successors of the regular limit cardinals. Thus the first is $\aleph_1$ and the next (if it exists) is the successor of the least inaccessible.

On steady unsurpassables, we have:

Lemma 2.   (i) No successor $\mathbf{u}^+$ of a steady cardinal $\mathbf{u}$ is steady;
            (ii) no limit cardinal $\mathbf{u}$ is steady;
            (iii) for any steady unsurpassable $\mathbf{u}$,
                $\mathrm{cp}(\mathbf{u}) \leq \mathrm{exbd}(\mathbf{u}) < \mathbf{u}$.

Proof. (i) $\mathrm{cp}(\mathbf{u}) = \mathrm{cp}(\mathbf{u}^+)$; and hence $2^{\mathrm{cp}(\mathbf{u}^+)} = 2^{\mathrm{cp}(\mathbf{u})} = \mathbf{u} < \mathbf{u}^+$.

(ii) Suppose $\mathbf{u}$ is steady limit cardinal. Then:
$$\mathbf{u}^{\mathrm{cf}(\mathbf{u})} \leq \mathbf{u}^{\mathrm{cp}(\mathbf{u})}, \text{ since for limit cardinals } \mathrm{cf}(\mathbf{u}) \leq \mathrm{cp}(\mathbf{u}),$$
$$= (2^{\mathrm{cp}(\mathbf{u})})^{\mathrm{cp}(\mathbf{u})}, \text{ since } \mathbf{u} \text{ is steady,}$$
$$= 2^{\mathrm{cp}(\mathbf{u}) \cdot \mathrm{cp}(\mathbf{u})}, \text{ by cardinal arithmetic,}$$
$$= 2^{\mathrm{cp}(\mathbf{u})}$$
$$= \mathbf{u}.$$
But $\mathbf{u}^{\mathrm{cf}(\mathbf{u})} > \mathbf{u}$ by cardinal arithmetic. A contradiction.

(iii) $\mathrm{cp}(\mathbf{u})$ is exponentially small (relative to $\mathbf{u}$) – for $\mathbf{u}^{\mathrm{cp}(\mathbf{u})} = (2^{\mathrm{cp}(\mathbf{u})})^{\mathrm{cp}(\mathbf{u})} = 2^{\mathrm{cp}(\mathbf{u})} = \mathbf{u}$; and so $\mathrm{cp}(\mathbf{u}) \leq \mathrm{exbd}(\mathbf{u})$. By (ii), $\mathbf{u}$ is a successor cardinal; and so if $\mathrm{exbd}(\mathbf{u}) = \mathbf{u}, \mathbf{u}$ itself must be exponentially small. But this contradicts Cantor's theorem.

Given a cardinal $\mathbf{c}$, we let $\mathbf{u_c}$ (or $\mathbf{u}/\mathbf{c}$) be the smallest unsurpassable greater than or equal to $\mathbf{c}$ (assuming that such a cardinal exists). Thus $\mathbf{u}_0$ is the smallest unsurpassable cardinal and hence, by Lemma 1 (ii), is steady.

If $\mathbf{u_c}$ exists, it may be constructed 'from below' in a natural manner. For each ordinal ξ, define the cardinals $\mathbf{u}_{\mathbf{c}, \xi}$ by:

(i) $\mathbf{u}_{\mathbf{c}, 0} = \mathbf{c}$;
(ii) $\mathbf{u}_{\mathbf{c}, \alpha+1} = \max(\mathbf{u}_{\mathbf{c}, \alpha}, 2^{\mathrm{cp}(\mathbf{u}_{\mathbf{c}, \alpha})})$; and
(iii) $\mathbf{u}_{\mathbf{c}, \lambda} = \lim_{\zeta < \lambda} \mathbf{u}_{\mathbf{c}, \zeta}$.

It is clear that $\mathbf{u}_{\mathbf{c}, \xi} \leq \mathbf{u_c}$ for each ξ. Moreover, $\mathbf{u}_{\mathbf{c}, \alpha} \leq \mathbf{u}_{\mathbf{c}, \beta}$ for any $\alpha \leq \beta$. On the assumption that $\mathbf{u_c}$ exists, there will therefore be an $\alpha \leq \mathbf{u_c}$ for which $\mathbf{u}_{\mathbf{c}, \alpha} = \mathbf{u}_{\mathbf{c}, \alpha+1} = \mathbf{u_c}$. As long as $\mathbf{c}$ is not an unsteady unsurpassable cardinal, we will always have $2^{\mathrm{cp}(\mathbf{u}_{\mathbf{c}, \alpha})} \geq \mathbf{u}_{\mathbf{c}, \alpha}$; and hence clause (ii) can be written in the simpler form:

(ii)' $\mathbf{u}_{\mathbf{c}, \alpha+1} = 2^{\mathrm{cp}(\mathbf{u}_{\mathbf{c}, \alpha})}$.

We conclude with a result on exponential smallness. Say that a cardinal $\mathbf{p}$ is *exponential* if it is of the form $2^{\mathbf{q}}$ for some $\mathbf{q}$. If $\mathbf{k}$ is exponentially small relative to $\mathbf{m}$, i.e. if $\mathbf{m}^{\mathbf{k}} \leq \mathbf{m}$, then $\mathbf{k}$ may not be exponentially small relative to a greater cardinal $\mathbf{p}$. However, this result will hold in case $\mathbf{p}$ is exponential:

Lemma 3. If $\mathbf{m}^k \leq \mathbf{m}$, then $\mathbf{p}^k \leq \mathbf{p}$ for $\mathbf{p}$ an exponential cardinal $\geq \mathbf{m}$.

Proof. The result is evident in case $\mathbf{m}$ is finite or $\mathbf{m} = \mathbf{p}$. So let us assume that $\mathbf{m}$ is infinite and that $\mathbf{p} = 2^{\mathbf{q}} > \mathbf{m}$. Now $\mathbf{q} \geq \mathbf{k}$. For assume otherwise, i.e. $\mathbf{k} > \mathbf{q}$. Then $\mathbf{m} \geq \mathbf{m}^k \geq 2^k \geq 2^{\mathbf{q}} = \mathbf{p}$, contrary to assumption. But then $\mathbf{p}^k = (2^{\mathbf{q}})^k = 2^{\mathbf{q} \cdot k} = 2^{\mathbf{q}} = \mathbf{p}$.

We now embark on the construction of the domain of the intended model. This proceeds by adding an abstract to a given domain if it can be produced by applying an acceptable means of abstraction to a concept that is defined over the given domain. In our account of acceptability, we need to modify the condition of being non-inflationary slightly in order to anticipate the possibility that the domain will contain two objects even if it does not already. Accordingly, let us say that an equivalence $\approx$ on $\wp(M)$ is *non-inflating* if $\mathrm{card}(P_{\approx}) \leq \max\{2, \mathrm{card}(M)\}$. Thus it is only in case $\mathrm{card}(M) < 2$ that there can be non-inflating equivalences that are not also non-inflationary. We now say that the equivalence relation $\approx$ on $\wp(M)$ is *K-acceptable* if $\approx$ is non-inflating and internally K-invariant for K a modest and exponentially small subset of M, that it is (*broadly*) *acceptable* if it is K-acceptable for some K, and that it is *strictly acceptable* if it is ϕ-acceptable. The strictly acceptable equivalences correspond to those acceptable means of abstraction that can be given a purely logical definition. (For $\mathrm{card}(M) \leq 1$, the definitions differ slightly from those previously given; and since we are only interested in internal invariance, we have dropped the I-suffix.)

We shall identify abstracts with cells. However, it will be important to indicate the domain from which the cell originated. Accordingly, we take an *indexed cell* to be an ordered pair $< X, M >$ for X a non-empty subset of $\wp(M)$. An indexed cell $< X, M >$ can be legitimately introduced on the basis of the domain M in the following ways: it is *K-admissable* if X is induced by some K-acceptable equivalence $\approx$ on $\wp(M)$; it is *broadly admissible* (or *admissible, simpliciter*) if it is K-admissible for some subset K of M; and it is (*strictly*) *admissible* if it is ϕ-admissible. It should be noted that, by Theorem III. 8.8, the admissible cells $< X, M >$ may be characterized, without reference to an underlying equivalence, as those for which X is predominantly I-invariant.

There is, however, a difficulty in representing an abstract x by means of an indexed cell $<X, M>$. For under an expansion M^+ of the domain, the abstract x might be associated with further concepts from $\wp(M^+)$. We therefore need a method for determining which concepts will be associated with the cell on the basis of the concepts that are already associated with it.

One obvious solution to this problem is to represent an abstract, not by a single cell, but by a sequence of cells, one for each of the expansions of the original domain. However, the details of this approach are extremely messy; and we prefer instead to fix upon a canonical way in which a cell can grow from one domain to another. The cell is taken to contain the 'seed' from which its subsequent growth is determined. We describe one natural way in which this can be done—though there are other possibilities, all giving rise to the same end-result, that might also be considered.

Suppose that M^+ is a superset of M. Then given an equivalence $\approx$ on $\wp(M)$, we call $\approx'$ an *expansion of* $\approx$ *from* M *to* M^+ if it is an equivalence on $\wp(M^+)$ that coincides with $\approx$ on $\wp(M)$; and given an indexed cell $<X, M>$, we say that the indexed cell $<X^+, M^+>$ is an *expansion of* $<X, M>$ if $X^+ \supseteq X$ and both X^+ and X agree on subsets of M. The cell $<X^+, M^+>$ is said to be a *K-legitimate* expansion of $<X, M>$ if X is induced by some K-acceptable equivalence $\approx$ on $\wp(M)$ and X^+ is induced by some K-acceptable expansion $\approx'$ of $\approx$ on $\wp(M^+)$. The expansion $<X, M>$ is *broadly legitimate* (or *legitimate, simpliciter*) if it is K-legitimate for some K; and it is *strictly legitimate* if it is ϕ-legitimate.

The legitimate expansions of a cell correspond to the different possible ways the abstract represented by the cell could behave in an extended domain. Given a set C (corresponding to a given concept), we must now somehow determine whether C should be associated with the cell $<X, M>$—or, as we shall put it, whether the cell *admits* the set—on the basis of the legitimate expansions $<X^+, M^+>$ of the cell. We make two decisions in this regard: we look only at the expansions of the form $<X^+, M \cup C>$; and we require that the set C should belong to every such expansion. Thus we say that $<X, M>$ *K-admits* the set C if (a) there is a K-legitimate expansion of $<X, M>$ of the form $<X^+, M \cup C>$ and (b) $C \in X^+$ for any K-legitimate expansion of the form $<X^+, M \cup C>$; and similarly for the definitions of *broad* and *strict admittance.*

We are able to give an intrinsic characterization of admittance. But first we need some results on the expansions of equivalences.

Lemma 4. Suppose that $\approx$ is a K-acceptable equivalence on $\wp(M)$ for K an exponentially small subset of M; and let N be a superset of M. There is then a K-legitimate expansion $\approx'$ of $\approx$ to N as long as M is finite or N is of the same cardinality as M or N is of exponential cardinality.

Proof. We distinguish between the finite and infinite case:

(i) M finite. Then K is empty and $\approx$ is strictly invariant. So it follows from Theorem III. 7.14 that $\approx$ is a numeric equivalence (with associated equivalence $\approx_{\text{num}}$ on the cardinals). Let us define $\approx'$ on subsets of N by:

$C \approx' D$ iff card(C) $\approx_{\text{num}}$ card(D).

Then it is readily verified that $\approx'$ is non-inflationary, internally invariant, and an expansion of $\approx$.

(ii) M infinite. Set $\mathbf{m} = \text{card}(M)$, and define $\approx'$ on subsets C and D of N by:

$C \approx' D$ iff either $\text{Icdstr}_K(C, D) \in \text{Icdstr}_K(\approx)$

or $\text{Icdstr}_K(C, C), \text{Icdstr}_K(D, D) \notin \text{Icdstr}_K(\approx)$.

Note that the condition $\text{Icdstr}_K(C, C) \in \text{Icdstr}_K(\approx)$ is satisfied just in case card(C) $\leq \mathbf{m}$.

It is readily verified that $\approx'$ is internally K-invariant. We show that $\approx'$ is an equivalence. It is clearly reflexive and symmetric. In order to establish transitivity, suppose that $C \approx' D$ and $D \approx' E$. If any of C, D, or E have cardinality $> \mathbf{m}$, then they all do; and so it is clear that $C \approx' E$. So suppose that they all have cardinality $\leq \mathbf{m}$. By thinning out members of M from C, D, and E, we may find sets C', D', and E' for which $M - (C' \cup D' \cup E')$ is of cardinality $\mathbf{m}$ and yet $\text{Icdstr}_K(C', D') = \text{Icdstr}_K(C, D), \text{Icdstr}_K(D', E') = \text{Icdstr}_K(D, E)$, and $\text{Icdstr}_K (C', E') = \text{Icdstr}_K(C, E)$. By the K-invariance of $\approx'$, $C' \approx' D'$ and $D' \approx' E'$. Let $B = (C' \cup D' \cup E') - M$. Choose a subset B^* of $M - (C' \cup D' \cup E' \cup K)$ of the same cardinality as B; and let $C^* = (C' - B) \cup B^*, D^* = (D' - B) \cup B^*$, and $E^* = (E' - B) \cup B^*$. It is readily verified that $\text{Icdstr}_K(C', D') = \text{Icdstr}_K(C^*, D^*), \text{Icdstr}_K (D', E') = \text{Icdstr}_K(D^*, E^*)$, and $\text{Icdstr}_K(C', E') = \text{Icdstr}_K(C^*, E^*)$. But then $C^* \approx D^*, D^* \approx E^*$. Hence $C^* \approx E^*$ by the transitivity of $\approx$; so $C' \approx' E'$ by the definition of $\approx'$; and so $C \approx' E$ by the K-invariance of $\approx'$.

To show that $\approx'$ is non-inflationary it suffices, by Corollary III. 7.16, to show that it contains the equivalence $\approx_{1,K}$ as defined on N. So suppose $C \approx_{1,K} D$, where C and D are exponentially large in N and of the same cardinality $\mathbf{p}$. There are two cases. (i) $\mathbf{p} > \mathbf{m}$. Then $C \approx' D$ by the definition of $\approx'$. (ii) $\mathbf{p} \leq \mathbf{m}$. By Lemma 3 above, $\mathbf{p}$ is exponentially large relative to card (M); and so $\approx$ will relate any two subsets of M of cardinality $\mathbf{p}$. But then we may readily find subsets C' and D' of M of cardinality $\mathbf{p}$ for which $\mathrm{Icdstr}_K(C', D') = \mathrm{Icdstr}_K(C, D)$; and so $C' \approx' D'$.

It should be noted that the expansion $\approx'$ is defined differently for the finite and for the infinite case. Neither definition will work for the other case. Suppose, for example, that we apply the definition for the infinite case to the finite case. Take a domain M with three elements x, y, z, and an acceptable equivalence $\approx$ that relates any two membered subsets of M. Then $\approx'$ will not even be an equivalence on any proper superset N of M. For take an element w in N but not in M. We will then have $\{x, y\} \approx' \{y, z\}$ and $\{y, z\} \approx' \{z, w\}$, but not $\{x, y\} \approx' \{z, w\}$.

Let us use $\approx_+$ and $\approx_K$, respectively, for the expansion $\approx'$ as defined in the finite and in the infinite case, but not restricted to any particular domain. Thus for *any* sets C and D:

$C \approx_+ D$ iff card(C) $\approx_{\mathrm{num}}$ card(D); and
$C \approx_K D$ iff either $\mathrm{Icdstr}_K(C, D) \in \mathrm{Icdstr}_K(\approx)$ or $\mathrm{Icdstr}_K(C, C)$, $\mathrm{Icdstr}_K(D, D) \notin \mathrm{Icdstr}_K(\approx)$.

We now characterize admittance, dealing first with the finite and then with the infinite case:

Lemma 5. For M finite, $\approx$ an acceptable equivalence on $\wp(M)$, X a cell induced by $\approx$, and $B \in X$, the following conditions are equivalent:

  (i) $<X, M>$ admits C;
  (ii) $C \approx_+ B$;
  (iii) card(C) = card(D) for some $D \in X$.

Proof. (i) implies (ii). Suppose $<X, M>$ admits C. Then some acceptable equivalence on $\wp(M)$ induces the cell X. By the proof of Lemma 4, the restriction of $\approx_+$ to $N = M \cup C$ is an acceptable equivalence on $\wp(N)$. But then, by the definition of admittance, $C \approx_+ B$.

(ii) implies (iii). Suppose $C \approx_+ B$. Then for some subsets D and E of M, $D \approx E$ with $\mathrm{card}(C) = \mathrm{card}(D)$ and $\mathrm{card}(B) = \mathrm{card}(E)$. By Theorem III. 7.14, the equivalence $\approx$ is numeric; and so $E \approx B$. Given $D \approx E$ and $D \approx B, D \approx B$; and so $D \in X$.

(iii) implies (i). Suppose $\mathrm{card}(C) = \mathrm{card}(D)$ for some D in X. Then $\mathrm{card}(C)$ is finite and hence so is $N = M \cup C$. By Lemma 4, there is an acceptable expansion of $\approx$ to $\wp(N)$. Take now any acceptable expansion $\approx'$ of $\approx$ to $\wp(N)$. By Theorem III.7.14, $\approx'$ is numeric; and so $D \approx' C$.

Lemma 6. For M infinite, $\approx$ a K-acceptable equivalence on $\wp(M), X$ a cell induced by $\approx$, and $B \in X$, the following conditions are equivalent:

(i) $<X, M>$ K-admits C;
(ii) $C \approx_K B$ (i.e. $\mathrm{Icdstr}_K(C, B) \in \mathrm{Icdstr}_K(\approx)$);
(iii) $\mathrm{Icdstr}_K(C, B) \in \mathrm{Icdstr}_K(X)$.

Proof. (i) $\rightarrow$ (ii). Suppose $<X, M>$ K-admits C. Then some K-acceptable equivalence $\approx'$ on $\wp(N)$, for $N = M \cup C$, induces an expansion $<Y, N>$ of $<X, M>$ that contains C. Now $\mathrm{card}(C) \leq \mathrm{card}(M)$. For suppose $\mathrm{card}(C) = \mathbf{n} > \mathrm{card}(M)$, and define a new equivalence on $\wp(N)$ by:

$D \approx'' E$ iff either $\mathrm{card}(D) = \mathrm{card}(E) = \mathbf{n}$ or $\mathrm{card}(D), \mathrm{card}(E) \neq \mathbf{n}$ and $D \approx' E$.

Then it is readily verified that $\approx''$ is K-acceptable and that it induces an expansion $<Y, N>$ of $<X, M>$ that does not contain C.

Now since $\mathrm{card}(N) = \mathrm{card}(M)$, it follows by the proof of Lemma 4 and the definition of K-admittance that the restriction $\approx^*$ of $\approx_K$ to $\wp(N)$ is K-acceptable. But then $C \approx^* B$; and hence $C \approx_K B$.

(ii) $\rightarrow$ (iii). Suppose $C \approx_K B$, i.e. $\mathrm{Icdstr}_K(C, B) = \tau = \mathrm{Icdstr}_K (E, F)$ for some E and F for which $E \approx F$. If B is exponentially large (in M) then, by Corollary III.7.16, we may 'thin' B and thereby obtain a B' for which $B \approx B'$, $\mathrm{Icdstr}_K(C, B') = \tau$ and $M - (B' \cup K)$ is large (in the non-exponential sense). Clearly it then suffices to prove the result by setting $B = B'$. If B is small, then $M - (B \cup K)$ is already large. So in either case, we may assume that $M - (B \cup K)$ is large.

Given that $\mathrm{Icdstr}_K(C, B) = \tau$, we can now find a subset C' of M for which $\mathrm{Icdstr}_K(C', B) = \mathrm{Icdstr}_K(C, B) = \tau$. So $C' \approx B$; and hence $\mathrm{Icdstr}_K(C, B) \in \mathrm{Icdstr}_K(X)$.

(iii) $\rightarrow$ (ii). Trivial.

(ii) → (i). Suppose that $\mathrm{Icdstr}_K(C, B) \in \mathrm{Icdstr}_K(\approx)$, i.e. $\mathrm{Icdstr}_K(C, B) = \mathrm{Icdstr}_K(E, F)$ for $E \approx F$. Given that $<X, M>$ is K-admissable, there is a K-acceptable equivalence $\approx$ on $\wp(M)$ that induces X. Given that $\mathrm{Icdstr}_K(C, B) \in \mathrm{Icdstr}_K(\approx)$, card(C) $\leq$ card(M). So card(N) = card(M) (for N = M $\cup$ C). But then, by Lemma 4, the restriction of $\approx_K$ to N is a K-legitimate expansion of $\approx$. Take now any K-legitimate expansion $\approx'$ of $\approx$ on $\wp(N)$. Since $E \approx F$, $E \approx' F$; and since $\mathrm{Icdstr}_K(C, B) = \mathrm{Icdstr}_K(E, F)$, $C \approx' B$. But then $C \in |B|_{\approx'} \supseteq X$.

What $<X, M>$ broadly or strictly admits can depend upon M. For suppose that $X = \{\{x\}\}$, M_1 is any finite set containing the doubleton $\{x, y\}$, and M_2 is any infinite set containing $\{x, y\}$. Then $<X, M_1>$ admits any singleton subset of M_1 while $<X, M_2>$ admits only the set $\{x\}$ itself. However, note that condition (ii) (or condition (iii)) in both Lemmas 5 and 6 makes no reference to M. This means that it is only the status of M as finite or infinite that can make a difference to what $<X, M>$ admits. We call $<X, M>$ *finitary* if M if finite and *infinitary* otherwise.

Given the two lemmas, we may prove:

Lemma 7. Suppose that the cell $<X, M>$ is admissible. Then:

(i) If it admits C, it L-admits C for any L for which $<X, M>$ is L-admissible;

(ii) It admits a subset C of M iff $C \in X$.

Proof. (i) There are two cases:

(a) M finite. The result is then trivial since only ϕ is exponentially small.

(b) M infinite. Suppose that $<X, M>$ admits C, i.e. K-admits C for K exponentially small. So $<X, M>$ is K-admissible and, by Lemma 6 (iii), $\tau = \mathrm{Icdstr}_K(C, B) \in \mathrm{Icdstr}_K(X)$ for some B in X. We may suppose that $M - B$ is large. For assume otherwise. Then B is large; and so, by Corollary III.7.16, X must contain any set B' that is equinumerous with B and agrees with B on K. But then, given that K is small, we may, by appropriate 'thinning' of B, find a B' in X for which $M - B'$ is large and $\mathrm{Icdstr}_K(C, B) = \mathrm{Icdstr}_K(C, B')$.

Let $D = C - M$. Note that in order for $\tau \in \mathrm{Icdstr}_K(X)$ we must have card(D) $\leq$ card(M). We distinguish two subcases. (1) card$(C - (B \cup D)) \geq$ card(D). We then let $C' = C - D$. (2) card$(C - (B \cup D)) <$ card(D). We then pick a subset of D' of

$M - (B \cup K \cup L)$ of the same size as D and let $C' = (C - D) \cup D'$. It is then readily ascertained, in either case, that $\mathrm{Icdstr}_{K \cup L}(C', B)$ $= \mathrm{Icdstr}_{K \cup L}(C, B)$. Since $\mathrm{Icdstr}_K(C', B) = \tau \in X$, $C' \in X$; and since $\mathrm{Icdstr}_L(C', B) = \mathrm{Icdstr}_L(C, B)$, X L-admits C by Lemma 6.

(ii) Again there are two cases:

(a) M finite. Given that $< X, M >$ admits a subset C of M, $\mathrm{card}(C) = \mathrm{card}(B)$ for some $B \in X$ by Lemma 5. So since any acceptable equivalence on $\wp(M)$ must be numeric, $C \in X$. Now suppose that $C \in X$. Then trivially C is of the same cardinality as some set in X; and so again by Lemma 5, $< X, M >$ admits C.

(b) M infinite. Assume that $< X, M >$ is admissible. Then, for some K exponentially small in M and some K-acceptable equivalence $\approx$ on $\wp(M)$, $X \in P_\approx$. Suppose now that $< X, M >$ admits the subset C of M. Then, for some sets D, E, F in X, $\mathrm{Icdstr}_K(C, D) = \mathrm{Icdstr}_K(E, F)$. So $D \approx E \approx F$; and, by $\approx$ internally K-invariant, $C \approx D$. But then $C \in X$. Next suppose that $C \in X$. Then $\mathrm{Icdstr}_K(C, C) \in \mathrm{Icdstr}_K(X)$; and hence X admits C in M.

Two indexed cells $< X, M >$ and $< Y, N >$ are said to be *indistinguishable* if they admit the same subsets of $M \cup N$ and to be *absolutely indistinguishable* if they admit all the same sets. Similar notions can be defined with strict in place of broad admittance. We establish two sufficient conditions for absolute indistinguishability.

Lemma 8. Suppose that the cells $< X, M >$ and $< Y, N >$ are admissible, $M \subseteq N$, and either M is finite or N is of the same cardinality as M or N is of exponential cardinality. Then the two cells are indistinguishable iff they are absolutely indistinguishable.

Proof. The direction from right to left is trivial. Suppose now that $< X, M >$ and $< Y, N >$ are indistinguishable. The case in which M is finite is straightforward. So let us suppose that M is infinite. Given that $< X, M >$ is admissible, X is induced by some K-acceptable equivalence $\approx$ on $\wp(M)$, for K an exponentially small subset of M. Take any subset C of N. Then:

$C \in Y$ iff Y admits C, by Lemma 7(ii)
    iff X admits C, by supposition
    iff $C \approx_K B$ for some $B \in X$.

By the proof of Lemma 4 and the conditions on N, it follows that the restriction $\approx'$ of $\approx_K$ to $\wp(N)$ is an acceptable equivalence that induces Y. Take now any C whatever. Then, by Lemma 6, $< Y, N >$

admits C iff $C \approx_K B$; and, by another application of Lemma 6, $C \approx_K B$ iff $<X, M>$ admits C.

Lemma 9. Suppose that $<X, M>$ and $<Y, N>$ are infinitary K-legitimate cells. Then they are absolutely indistinguishable if X and Y have a member in common and $\mathrm{Icdstr}_K(X) = \mathrm{Icdstr}_K(Y)$.

Proof. Let B be the common member of X and Y. Suppose that $<X, M>$ admits C. Then by Lemma 7(i), it K-admits C. So by Lemma 6, $\mathrm{Icdstr}_K(C, B) \in \mathrm{Icdstr}_K(X)$. So $\mathrm{Icdstr}_K(C, B) \in \mathrm{Idstr}_K(Y)$; and hence, by Lemma 6 again, $<Y, L>$ K-admits C.

The construction of the domain U can now be described. Let I be a set of urelements (which we regard as the individuals, i.e. the non-abstracts). We make no assumption regarding I beyond the fact that $\mathrm{card}(I)$ is not an unsteady unsurpassable. The construction then proceeds relative to a fixed choice of I. For each ordinal ξ, we define $U_{I,\xi}$ by:

(i) $U_{I,0} = I$;

(ii) $U_{I,\alpha+1} = U_{I,\alpha} \cup \{ <X, U_{I,\alpha}> : <X, U_{I,\alpha}>$ is an admissible cell distinguishable from every cell in $U_{I,\alpha}\}$;

(iii) $U_{I,\lambda} = \bigcup_{\zeta < \lambda} U_{I,\zeta}$.

For the purposes of the strict construction, the Us should be changed to Vs and clause (ii) should be replaced with:

(ii)′ $V_{I,\alpha+1} = V_{I,\alpha} \cup \{ <X, V_{I,\alpha}> : <X, V_{I,\alpha}>$ is a strictly admissible cell and is strictly distinguishable from every cell in $V_{I,\alpha}\}$.

We say that the cell $<X, M>$ is *introduced at* the ordinal α if $<X, M>$ belongs to $U_{I,\alpha}$ but to no $U_{I,\beta}$ for $\beta < \alpha$ (and similarly for the strict construction).

We need to know that the construction will stabilize, i.e. that $U_{I,\alpha} = U_{I,\alpha+1}$ for some ordinal α. To this end, we compute the cardinalities of the domains U_ξ.

Lemma 10. (i) $\mathrm{card}(U_{I,\alpha}) \leq \mathrm{card}(U_{I,\beta})$ for $\alpha \leq \beta$;

(ii) $\mathrm{card}(U_{I,\alpha+1}) \geq 2^{\mathrm{cp}(\mathrm{card}(U_{I,\alpha}))}$ for $U_{I,\alpha}$ infinite;

(iii) if $U_{I,\alpha}$ is finite, then $\mathrm{card}(U_{I,\alpha+1})$ is finite and $> \mathrm{card}(U_{I,\alpha})$;

(iv) if $U_{I,\alpha}$ is infinite, then $\mathrm{card}(U_{I,\alpha+1}) \leq \max(\mathrm{card}(U_{I,\alpha}), 2^{\mathrm{cp}(\mathrm{card}(U_{I,\alpha}))})$;

(v) $\mathrm{card}(U_{I,\lambda}) = \lim_{\xi < \lambda} \mathrm{card}(U_{I,\zeta})$.

Proof.(i) Evident from the definition.

(ii) Set $\mathbf{u} = \text{card}(U_{I,\alpha})$; and let Γ be a non-empty set of cardinals $\leq \text{cp}(\mathbf{u})$. Define $\equiv_\Gamma$ on $U_{I,\alpha}$ by: $C \equiv_\Gamma D$ iff card(C) and card(D) both belong to Γ or neither do. Then the equivalence $\equiv_\Gamma$ is clearly internally invariant; and since the associated partition contains at most two equivalence classes, it is non-inflating on $U_{I,\alpha}$. So if X_Γ is the equivalence class of the form $|C|$ for card(C) $\epsilon \ \Gamma$, $<X_\Gamma, U_{I,\alpha}>$ is admissible. For distinct Γ, the X_Γ's are distinct; and so, given that $\mathbf{u}$ is transfinite, there are $2^{\text{cp}(\mathbf{u})}$ such cells.

(iii) Let $n = \text{card}(U_{I,\alpha+1})$ and $m = \text{card}(U_{I,\alpha})$. Then $n \leq 2^p$, for $p = 2^m$, and hence is finite given that m is finite. We see that $n > m$ from the construction of $\equiv_\Gamma$ above. For this shows that $n \geq (2^{m+1} - 1) > m$.

(iv) Fix on a particular exponentially small subset K of $U_{I,\alpha}$ of cardinality $\mathbf{k} < \mathbf{u} = \text{card}(U_{I,\alpha})$. Then, by Theorem III.7.5, the number of internally K-invariant cells over $U_{I,\alpha+1}$ is at most $\max(\mathbf{u}, 2^{\text{cp}(\mathbf{u})}, 2^{\mathbf{k}}) = \max(\mathbf{u}, 2^{\text{cp}(\mathbf{u})})$, given that $\mathbf{k}$ is exponentially small. But there are at most $\sup\{\mathbf{u}^{\mathbf{k}}:\mathbf{k} \text{ exponentially small}\} \leq \mathbf{u}$ such K, s; and hence card$(U_{I,\alpha+1}) \leq \max(\mathbf{u}, 2^{\text{cp}(\mathbf{u})})$.

(v) Evident from the definition.

Let us note that the proofs of the lower bounds (namely (ii) and (iii)) appeal only to the strictly invariant equivalences $\equiv_\Gamma$. Thus the results will also hold for the strict construction $V_{I,\alpha}$ in place of $U_{I,\alpha}$. If $\mathbf{c} = \text{card}(I)$ is infinite, it follows from (i), (ii), (iv), and (v), that card$(U_{I,\alpha}) = \mathbf{u}_{\mathbf{c},\alpha}$. If $\mathbf{c}$ is finite, it will not be generally true that card$(U_{I,\alpha}) = \mathbf{u}_{\mathbf{c},\alpha}$. However, the truth of the identity will be restored once we pass to infinite α.

Assuming the existence of the unsurpassable cardinal $\mathbf{u}_\mathbf{c}$, we can show that the constructions will stabilize.

Lemma 11. The constructions $U_{I,\alpha}$ and $V_{I,\alpha}$ stabilize on the ordinal $\mathbf{u}_\mathbf{c} + \mathbf{u}_\mathbf{c}$ (given that $\mathbf{c} = \text{card}(I)$ is not an unsteady unsurpassable).

Proof. We deal with the case of $U_{I,\alpha}$, the other case being similar. It follows from Lemma 10 above that card$(U_{I,\mathbf{u}_\mathbf{c}}) = \mathbf{u}_\mathbf{c}$. We need to show that no new cells are introduced at the ordinal stage $\mathbf{u}_\mathbf{c} + \mathbf{u}_\mathbf{c}$. To simplify notation, set $\mathbf{u} = \mathbf{u}_\mathbf{c}$ and $W_\alpha = U_{I,\mathbf{u}+\alpha}$, and note that card$(W_\mathbf{u}) = \mathbf{u} = \text{card}(W_0)$.

We first establish:

(1) Any exponentially small subset K of $W_\mathbf{u}$ is included in W_α for some $\alpha < \mathbf{u}$.

Proof. Suppose the condition is not true for some K. Then for each $\alpha < \mathbf{u}$, there is a β, with $\alpha < \beta < \mathbf{u}$, and a member of K that is introduced at β (in the sequence $< W_\xi : \xi < \mathbf{u} >$). So card$(K) \geq$ cf$(\mathbf{u})$. But by Lemma 2(ii), $\mathbf{u}$ is a successor cardinal. So cf$(\mathbf{u}) = \mathbf{u}$; and therefore K is not small after all.

From (1) we establish (2) below, from which the desired result follows:

(2) Any admissible cell $< X, W_\mathbf{u} >$ is indistinguishable from a cell in W_α for $\alpha < \mathbf{u}$.

Proof. It suffices to show that $< X, W_\mathbf{u} >$ is indistinguishable (and hence, by Lemma 8, is absolutely indistinguishable) from an admissible cell of the form $< Y, W_\alpha >$ for $\alpha < \mathbf{u}$. For then either $< Y, W_\alpha >$ is a member of $W_{\alpha+1}$ or it is indistinguishable (and hence absolute indistinguishable) from a member of W_α; and so, in either case, $< X, W_\mathbf{u} >$ is absolutely indistinguishable from a cell in W_α.

Suppose that $< X, W_\mathbf{u} >$ is K-admissible for K an exponentially small subset of $W_\mathbf{u}$. By (1), $K \subseteq W_\alpha$ for some $\alpha < \mathbf{u}$. We distinguish two cases:

(a) X contains no exponentially small set. For subsets C and D of $W_\mathbf{u}$, define C eq$_K$ D by: $C \cap K = D \cap K$ and card$(C{-}K)$ = card$(D{-}K)$. Note that, for exponentially large subsets C and D of $W_\mathbf{u}$, $C \approx_{1, K} D$ iff C eq$_K$ D. So by Corollary III.7.16, the cell X is closed under the relation eq$_K$. Let $Y = X \cap \wp(W_\alpha)$. Then it is readily verified that $< Y, W_\alpha >$ is admissible, that Icdstr$_K(Y)$ = Icdstr$_K(X)$, and that X and Y have a member in common. But then, from Lemma 8 above, it follows that the indexed cells are indistinguishable.

(b) X contains an exponentially small set C. By (1), $C \subseteq W_\beta$ for some β for which $\alpha \leq \beta < \mathbf{u}$. But now we may find an admissible cell $< Y, W_\beta >$ that is indistinguishable from $< X, W_\mathbf{u} >$. For let f be a one-to-one function from $W_\mathbf{u}$ onto W_β that is fixed over $C \cup K$; and let $Y = f[X]$. Then it is readily verified that $< Y, W_\beta >$ is K-admissible, that Icdstr$_K(Y)$ = Icdstr$_K(X)$, and that $C \in X, Y$. So by Lemma 8 again, $< X, W_\mathbf{u} >$ and $< Y, W_\beta >$ are indistinguishable.

We are now in a position to define the intended model $\mathbf{U}_I$. Its domain U_I is given by $U_{I, \alpha}$ for α a stabilizing ordinal. The interpretation *Abstr* of the predicate Abstr, for $x \in U_I, C \subseteq U_I$, and $R \subseteq \wp(U_I) \times \wp(U_I)$, is given by:

x $Abstr_R$ C iff R is an acceptable equivalence on $\wp(U_I)$ and $\{D: x$ admits $D\} = \{D: R$ relates C to $D\}$.

To obtain the strict model, we define V_I in analogy to U_I and define *Abstr* by:

x $Abstr_R$ C iff R is a strictly acceptable equivalence on $\wp(V_I)$ and $\{D: x$ strictly admits $D\} = \{D: R$ relates C to $D\}$.

It should be noted that there is a sense in which the construction of U_I serves to generate not only the abstracts from stage to stage but also the means of abstraction. Of course, an equivalence over the whole domain U_I cannot be constructed at a stage at which we have not yet constructed the whole domain. However, we may at each stage take ourselves to be constructing a *representative* of such an equivalence. For an equivalence $\approx$ over M can be taken to represent the restriction of $\approx_+$ to $\wp(U_I)$ when M is finite and to represent the restriction of $\approx_K$ to $\wp(U_I)$ when M is infinite and $\approx$ is K-acceptable. Each abstract can then be taken to be generated from a previously given concept by means of a representative of a means of abstraction that may also be taken to be previously given.

We may at last show that the model U_I verifies the axioms of GA (the proof that V_I is a model for SGA is similar).

Theorem 12. Suppose that card(I) is not an unsteady unsurpassable. Then U_I is a model of GA$^+$ in which each cell is an abstract and each abstract a cell.

Proof. It is clear that each abstract is a cell. To show the converse, let $<X, U_{I,\beta}>$ be a cell in U_I. We deal solely with the case in which $U_{I,\beta}$ is infinite, the other case, in which it is finite, being similar. Suppose that $<X, U_{I,\beta}> \in U = U_I$. Then $<X, U_{I,\beta}>$ is admissible. So, for some exponentially small subset K of $U_{I,\beta}$, there is a K- acceptable equivalence $\approx$ on $\wp(U_{I,\beta})$ that induces X. By Lemmas 1, 2(i), and 10, U is of exponential cardinality; and so, by Lemma 3, K is an exponentially small subset of U and the restriction $\approx'$ of $\approx_K$ to $\wp(U)$ is also an acceptable equivalence. Choose a B in X. By Lemma 6, $<X, U_{I,\beta}>$ admits a subset C of U iff $C \approx' B$; and so $<X, U_{I,\beta}>$ stands in the relation $Abstr_{\approx'}$ to B.

Let us now verify the axioms in turn:
Identity. It suffices to establish:

(*) any two distinct cells $<X, U_{I,\alpha}>$ and $<Y, U_{I,\beta}>$ of U_I, $\alpha \leq \beta$, are distinguishable.

Proof. Suppose that $\alpha = \beta$. Then there is a member C of one of the cells, say X, that is not a member of the other. By Lemma 7(ii), $<X, U_{I,\alpha}>$ admits C while $<Y, U_{I,\alpha}>$ does not; and hence the two indexed cells are distinguishable. On the other hand, if $\alpha < \beta$, then it follows from the construction that $<Y, U_{I,\beta}>$ is distinguishable from $<X, U_{I,\alpha}>$.

Broad Internal Existence. Let $\approx$ be an acceptable equivalence on $\wp(U_I)$. We need to show that, for every set $C \subseteq U_I$ (which represents a concept), there is a cell $<X, U_{I,\alpha}>$ in U_I (which represents an abstract) for which $\{D \subseteq U_I: <X, U_{I,\alpha}>$ admits D$\} = |C|_\approx$. Choose an α for which $U_{I,\alpha} = U_I$. By Lemma 7(ii), $\{D \subseteq U_I: <|C|_\approx, U_I>$ admits D$\} = |C|_\approx$. By the construction, $<|C|_\approx, U_I>$ is indistinguishable from a cell $<X, U_\beta>$ in U_I; so $\{D \subseteq U_I: <|C|_\approx, U_I>$ admits D$\} = \{D \subseteq U_I: <X, U_{I,\beta}>$ admits D$\}$; and so $\{D \subseteq U_I: <X, U_{I,\beta}>$ admits D$\} = |C|_\approx$.

Exclusivity and Application. Straightforward.

Strong Minimality. Suppose that there existed a proper subset D of U_I of the sort ruled out by Strong Minimality. Let us first show that D has the same cardinality as U_I. To this end, let Γ be a non-empty set of cardinals $\leq$ cp (D). Define $\approx\,=\,\equiv_\Gamma$ as in the proof of Lemma 10(ii), though relative now to the domain D; and let $\approx'$ be the restriction of $\approx_+$ to $\wp(U_I)$. Then clearly, $\approx'$ is strictly acceptable (and, given that card$(D) \geq 2$, non-inflationary and not just non-inflating). Let C_0 be any subset of D whose cardinality is in Γ. Then some object x in U_I is an abstract of C_0 with respect to $\approx'$; and so there is an abstract x_Γ in D for which $\{C \subseteq D: x_\Gamma$ is an abstract of $C\} = \{C \subseteq D:$ card$(C) \in \Gamma\}$. But the x_Γ are different for different Γ. This shows that card(D) $\geq 2^{\text{cp(card(D))}} - 1$ if D is finite and that card$(D) \geq 2^{\text{cp(card(D))}}$ if D is infinite; and from this it follows that card$(D) = $ card(U_I).

Suppose now that $D \neq U_I$. Let β be the least ordinal for which $D \subseteq U_{I,\beta}$ fails; and let x be an indexed cell in $U_{I,\beta}$ but not in D. So β is of the form $\alpha + 1$ and x is of the form $<X, U_{I,\alpha}>$ for $X \subseteq \wp(U_{I,\alpha})$. There is therefore an abstract x' in D that admits the same subsets of $D \subseteq U_{I,\alpha}$ as x. But then x' is indistinguishable from x; and hence $x = x'$, contrary to supposition.

Corollary 13. Let **c** be any cardinal that is not an unsteady unsurpassable. Then there is a standard model M of GTA containing exactly **c** individuals.

Proof. Choose I to be a set of urelements having cardinality **c**.

We state without proof:

Theorem 14. (Categoricity). Suppose that M and N are two standard models of GA^+ and that $\text{card}(I_M) = \text{card}(I_N)$ is not an unsteady unsurpassable. Then M and N are isomorphic.

3. Derivations

We briefly indicate how to derive the central portions of arithmetic and analysis within the theory of abstracts and within the corresponding theory of extensions, thereby providing those branches of mathematics with a neo-Fregean foundation.

In order to carry out the derivation within the theory of abstracts, we need the premiss that there are at least two objects (i.e. $\exists x, y(x \neq y)$). For Frege, this premiss follows from the fact that sentences are the names of truth-values; and it can perhaps therefore be deemed a logical truth. But under a more orthodox approach, the premiss must be be treated as additional (and presumably non-logical) axiom. Under the approach that uses extensions instead of abstracts, this additional axiom is not required.

The only other axioms we need within the theory of abstracts are Internal Existence (in its second-order version) and Identity in its left-to-right direction (forbidding the identification of abstracts that are associated with different equivalence classes of concepts). The only axioms we need within the theory of extensions are Extensionality and Comprehension in its strict schematic formulation. We shall merely provide an informal sketch of the proofs but with enough detail, I hope, to make clear how they might be formalized. We concentrate on the more problematic case of the derivations within the theory of abstracts.

Recall that a divisor divides the universe of concepts into at most two parts (thus Div (**R**) is defined by $\exists C, D \forall E(\mathbf{R}[C, E] \lor \mathbf{R}(D, E))$). Thus given that there are at least two objects, Internal Existence straightforwardly yields the conclusion that each internally I-invariant divisor is applicable, i.e. Eq(**R**) & Div(**R**) & I-inv(**R**) $\rightarrow$ App(**R**).

With each concept C, we may associate the strictly I-invariant divisor N_C, where $N_C(E, F) \leftrightarrow (E$ eq $C \leftrightarrow F$ eq $C)$ for any concepts E and F. Thus N_C divides the concepts into those that are equinumerous with C and those that are not. It should be clear that N_C is a divisor: for if every concept is equinumerous with C then every concept bears the relation N_C to C; while if some concept D is not equinumerous with C then every concept bears the relation N_C either to C or to D. Also, it is evident from the properties of equinumerosity that N_C is strictly I-invariant.

So it follows from Internal Existence that each of the divisors N_C is legitimate; and so, for each concept C, there is a unique abstract n_C of C with respect to the equivalence N_C. We could take n_C to be the cardinality of C. But it is also possible to derive Hume's Law and thereby see each cardinal as an abstraction from the same underlying equivalence relation.

To this end, we set up the correlation of each concept C with the abstract n_C. This correlation then establishes that equinumerosity is non-inflationary. For suppose that C and D are not equinumerous. Then n_C and n_D must be distinct; for their being the same would require that C be related by N_D to D. (Note that essential use is made here of the identity axiom.) Since equinumerosity is internally invariant, it follows that it is applicable; and hence Hume's Law is established.

We may also establish the legitimacy of a counterpart to sets of numbers. For let C be any non-empty concept of numbers. Then with C we may associate the divisor S_C, where $S_C(E, F) \leftrightarrow (Cn_E \leftrightarrow Cn_F)$. This divisor divides the universe of concepts into those whose number falls under C and those whose number falls outside of C. The divisor is readily shown to be internally invariant; and so for each concept C of numbers there exists the abstraction s_C of D wrt S_C, where D is some concept whose number n_D falls under C. The abstracts s_C may then be used to represent the non-empty sets of numbers. (It is not clear, however, how we might represent sets of numbers and, indeed, given the consistency of CH, there is no assurance that our models will have a cardinality greater than $2^{\aleph_0}$.)

It should be noted that the above proof makes no appeal to a principle of extensional abstraction, of either an unrestricted or a restricted sort. Indeed, the proof depends crucially upon the fact that the concepts in question are concepts of *numbers*. For it is their being numbers that entitles us to conclude that the corresponding equivalence relations are invariant.

We therefore see that a foundation can be provided for arithmetic and analysis without making any appeal to extensional abstracts. It is indeed possible to prove a form of extensional abstraction. For let us define $C \equiv^* D$ by: C and D are both exponentially small and coextensive or neither is exponentially small. Then it may be shown that $\equiv^*$ is strictly I-acceptable and hence it may be shown that each exponentially small concept has an extension. However, this result is of no help in deriving analysis. For in order to apply it to concepts of numbers we need to know that they are exponentially small; and this seems to require something like the earlier construction.

Within the theory of extensions, the derivations are much more straightforward. Hume's Law can be established directly, in much the same way as for Frege. For it is readily verified that, for each concept, the condition of being equinumerous with that concept is I-invariant and so, by Extensionality and Comprehension, it follows that there exists a unique extension whose members are the concepts that conform to the condition. We take that extension to be the number of the concept; and, with that understanding of number, it is a simple matter to derive Hume's Law.

In regard to the reals, we may, for each non-empty concept of numbers C, consider the condition of being a concept whose number falls under C. This condition is readily shown to be I-invariant and hence, for each such concept C, there will be an extension whose members are the concepts whose number falls under C. These extensions can then be used, as before, to represent the reals.

4. Further Work

Let me briefly mention three further topics that need to be considered.

(1) *Conservative Extensions.* Let us say that the §-theory $T^§$ (in the language $L^§$) is a *conservative extension of* the underlying second-order logic L^2 (with or without Well-Ordering) if the relativization ϕ^I of an §-free formula ϕ to the predicate I (for being an individual) is a theorem of $T^§$ only if ϕ is a theorem of L^2. Note that we do not require that every §-free theorem of $T^§$ be a theorem of L^2 since adopting an abstraction principle may result in our being able to prove that there are more objects than there were without the principle. It is plausible to suppose that conservativity, in this sense,

is a necessary condition for what, in sect. I. 1, I called a 'genuine' principle of abstraction; for we do not want such a principle to have any substantive implications for the logic of individuals. For this reason, and also for its intrinsic interest, it would be desirable to prove conservativity results for particular theories, such as the theory of numbers or restricted extensions, and also for general classes of theories. However, it is not clear to me how the proofs should go.[1]

(2) *Further Forms of Abstraction.* The theory of abstraction that we have developed is not comprehensive. For it does not deal with abstraction that is over objects, or over several concepts or relations at a time, or over the various combinations of these items. It would be desirable to extend the whole theory to these other cases. Perhaps the criterion of being non-inflationary and (predominantly) invariant could still be adopted as a means of avoiding the problem of hyper-inflation. But, in other respects, the theory will be quite different. We cannot expect to have the relational analogue of Hume's Law, for example. For, as Hodes (1984: 138) and Hazen (1985: 253–40) have pointed out, the abstraction principle for relation numbers will lead to the Burali-Forti paradox. Thus the possibilities for abstraction will be much more circumscribed than they are for concepts.

(3) *Alternative Forms of Second-Order Logic.* For many of our results, we have assumed that the underlying logic contains full second-order Comprehension for concepts and relations. It would be of interest to see how the theory of abstraction might be developed under various weakenings of this assumption. Some of our results also depend upon assuming that the underlying logic contains some version of the axiom of choice. But there are well-motivated theories of extension that are incompatible with such an as-sumption. For example, there is the Cantorian theory in which $\S C = \S D \leftrightarrow \forall x(Cx \leftrightarrow Dx)$ is taken to hold under the condition that C and D can be well-ordered. By using the reasoning of the paradoxes, it can be shown, within such a theory, that the universe cannot be well-ordered. Again, it would be of interest to see to what extent a theory of abstracts could be developed along these lines and whether, in particular, some of the difficulties over hyperinflation could thereby be avoided.

---

[1] There has recently been a negative result: Weir and Shapiro (1999) have shown that Boolos's new V is *not* conservative. The requirement of conservativity has also been discussed by Wright (1997: 221–5; 1999: sect. 5). He claims that the theory based upon Hume's Law *is* conservative but it is an open question, as far as I know, whether or not this is so.

References

Boolos, G. (1968), 'Degrees of Unsolvability of Constructible Sets of Integers', *Journal of Symbolic Logic*, 33/4: 497–513.

—— (1986–7), 'Saving Frege from Contradiction', *Proceedings of the Aristotelian Society*, NS 87: 137–51.

—— (1987), 'The Consistency of Frege's "Foundations of Arithmetic"', in J. J. Thomson (ed.), *On Being and Saying: Essays in Honor of Richard Cartwright* (Cambridge, Mass.: MIT Press), 3–20.

—— (1990), 'The Standard of Equality of Numbers', in id. (ed.), *Meaning and Method* (Cambridge: Cambridge University Press), 261–77.

—— (1993), 'Whence the Contradiction', *Proceedings of the Aristotelian Society*, suppl. vol. 67: 213–33.

—— (1997), *Is Hume's Principle Analytic*, in Heck (1997: 245–62).

Brandl, J., and Sullivan, P. (1998) (eds.), *New Essays on the Philosophy of Michael Dummett* (Vienna: Rodopi).

Cartwright, R. (1994), 'Speaking of Everything', *Nous*, 28: 1–20.

Drake, F. R. (1974), *Set Theory: An Introduction to Large Cardinals* (Amsterdam: North-Holland).

Dummett, M. A. E. (1963), 'The Philosophical Significance of Godel's Theorem', *Ratio*, 5: 140–55 (repr. in Dummett 1978: 186–201).

—— (1973), *Frege: Philosophy of Language* (London: Duckworth).

—— (1978), *Truth and Other Enigmas* (London: Duckworth).

—— (1981*a*), *Frege: Philosophy of Mathematics* (London: Duckworth).

—— (1981*b*), *The Interpretation of Frege's Philosophy* (London: Duckworth).

—— (1991*a*), *Frege: Philosophy of Mathematics* (London: Duckworth).

—— (1991*b*), *Frege and Other Philosophers* (Oxford: Clarendon Press).

—— (1993), *The Seas of Language* (Oxford: Clarendon Press).

—— (1998), *Neo-Fregeans: in Bad Company?*, in Schirn (1998: 243–51).

Field, H. (1984), 'Critical Notice of Crispin Wright's "Frege's Conception of Numbers as Objects"', *Canadian Journal of Philosophy*, 14: 637–62, repr. in Field (1989: ch. 5).

—— (1989), *Realism, Mathematics and Modality* (Oxford: Blackwell).

—— (2001), *Truth and the Absence of Fact* (Oxford: Oxford University Press).

Fine, K. (1989), 'The Problem of De Re Modality', in J. Almog, J. Perry, and H. Wettstein (eds.), *Themes From Kaplan* (Oxford: Clarendon Press), 197–272.

FINE, K. (1990), 'Quine on Quantifying In', in Anderson, and Owens (eds.), *Proceedings of the Conference on Propositional Attitudes* (Stanford, Calif.: CSLI).

—— (1994), 'Essence and Modality', *Philosophical Perspectives*, 8: 1–16.

—— (1995), 'Senses of Essence', in W. Sinnott-Armstrong (ed.), *Modality, Morality and Belief* (Cambridge: Cambridge University Press).

—— (1998), 'A Defence of Cantorian Abstraction', *Journal of Philosophy*, 98/12: 599–634.

FOREMAN, M., and WOODIN, W. H. (1991), 'The Generalized Continuum Hypothesis can Fail Everywhere', *Annals of Mathematics*, 133: 1–35.

FREGE, G. (1879), *Begriffsschrift* (Halle: L. Nebert). Trans. in part in Geach and Black (1952).

—— (1959), *Grundlagender Arithmetik*. Trans. J. L. Austin as *Foundations of Arithmetic* (Oxford: Blackwell).

—— (1992), 'Über Begriff under Gegenstand', *Viereljahrsschrift für Wissenschaftliche Philosophie*, 16: 192–205. Trans. as *Concept and Object* in Geach and Black (1952).

—— (1893–1903), *Grundesegetze der Arithmetik:* (Jena: H. Pohle). Trans. in part by M. Furth in *The Basic Laws of Arithmetic: Exposition of the System* (Berkeley: University of California Press).

GEACH, P., and BLACK, M. (1952), *Translations from the Philosophical Writings of Gottlob Frege* (Oxford: Blackwell).

HALE, B. (1987), *Abstract Objects* (Oxford: Blackwell).

—— (1994), 'Dummett's Critique of Wright's Attempt to Resuscitate Frege', *Philosophia Mathematica*, 2/3: 122–47; repr. in Hale and Wright 2000: ch. 8.

—— (1997), 'Grundlagen §64', *Proceedings of the Aristotelian Society*, 97: 243–61; repr. with a postscript in Hale and Wright 2000*a*: ch. 4.

—— (2000) 'Reals by Abstraction', *Philosophia Mathematica*, 3/8; repr. in Hale & Wright (2000*a*: ch. 15).

HALE, B., and WRIGHT, C. (2000*a*), *The Reason's Proper Study: Essays Towards a Neo-Fregean Philosophy of Mathematics* (Oxford: Clarendon Press).

—— (2000*b*), 'Implicit Definition and the A Priori', in P. Boghossian and C. Peacocke (eds.), *New Essays on the A Priori* (Oxford: Clarendon Press); repr. in Hale & Wright 2000*a*: ch. 5.

—— (2000*c*), 'To Bury Caesar', in Hale and Wright (2000*a*: ch. 14).

HAZEN, A. (1985), 'Review of Crispin Wright's "Frege's Conception of Numbers as Objects"', *Australasian Journal of Philosophy*, 63/2: 251–4.

HECK, R. G. (1992), 'On the Consistency of Second-Order Contextual Definitions', *Nous*, 64/4: 491–4.

—— (1993), 'The Development of Arithmetic in Frege's Grundgesetze der Arithmetik', *Journal of Symbolic Logic*, 58/2: 579–601.

—— (1996), 'Definition by Induction in Frege's *Grundgesetze der Arithmetik*', in M Schirn (ed.), *Frege: Importance and Legacy* (Berlin: Walter de Gruyter).

—— (1997), *Language, Thought and Logic* (Oxford: Oxford University Press).

HODES, H. T. (1984), 'Logicism and the Ontological Commitments of Arithmetic', *Journal of Philosophy*, 81/3: 123–49.

HORWICH, P., (1997), 'Implicit Definition, Analytic Truth and A Priori Knowledge', *Nous*, 31: 423–40.

—— (1998), *Meaning* (Oxford: Clarendon Press).

LINSKY, B. (1992), 'A Note on the "Carving up Content" Principle in Frege's Theory of Sense', *Notre Dame Journal of Formal Logic*, 33/1.

MENZEL, C. (1986), 'On the Iterative Explanation of the Paradoxes', *Philosophical Studies*, 49: 37–61.

PARSONS, C. (1964), 'Frege's Theory of Number', in M. Black (ed.), *Philosophy in America* (Ithaca, NY: Cornell University Press), 180–203.

—— (1997), 'Wright on Abstraction and Set Theory', in Heck (1997).

PARSONS, T. (1987), 'On the Consistency of the First-Order Portion of Frege's Logical System', *Notre Dame Journal of Formal Logic*, 28/1: 161–8.

PEACOCKE, C. (1991), 'The Metaphysics of Concepts', *Mind*, 100: 525–46.

ROSEN, G. (1993), 'The Refutation of Nominalism?', *Philosophical Topics*, 21: 149–86.

SCHIRN, M. (1998), *The Philosophy of Mathematics Today* (Oxford: Clarendon Press).

SCHROEDER-HEISTER, P. (1987), 'A Model-Theoretic Reconstruction of Frege's Permutation Argument', *Notre Dame Journal of Formal Logic*, 28/1: 69–79.

SHAPIRO, S. (1991), *Foundations without Foundationalism: A Case for Second-Order Logic* (Oxford: Clarendon Press).

SHAPIRO, S., and WEIR, A. (1999), 'New V, ZF and Abstraction', *Philosophia Mathematica*, 3/7: 293–321.

TENNANT, N. (1997), 'On the Necessary Existence of Numbers', *Nous*, 31: 30–6.

WRIGHT, C. (1983), *Frege's Conception of Numbers as Objects* (Aberdeen: Aberdeen University Press).

—— (1988), 'What Numbers can Believably Be', in *Revue Internationale de Philosophie*, 42: 425–73.

—— (1990), 'Field and Fregean Platonism', in A. D. Irvine (ed.), *Physicalism in Mathematics* (Dordrecht: Kluwer), 73–93; repr. in Hale and Wright (2000*a*: ch. 6).

—— (1997), 'On the Philosophical Significance of Frege's Theorem', in Heck (1997); repr. in Hale and Wright (2000*a*).

WRIGHT, C. (1998*a*), 'On the (Harmless) Impredicativity of Hume's Principle', in Schirn (1998: 339–68); repr. in Hale and Wright (2000*a*: ch. 10).

—— (1998*b*), 'Response to Dummett', in Schirn (1998: 389–405); repr. in Hale and Wright (2000*a*: ch. 11).

—— (1999), 'Is Hume's Principle Analytic?', *Notre Dame Journal of Formal Logic*, 40/1; repr. in Hale and Wright (2000*a*: ch. 13).

Main Index

abstracts
  versus abstractions 9
  access to 77
  axioms for, *see* abstracts – general
    theory of
  as equivalence classes 2, 14, 47, 49,
    52, 73–4
  existence of 7, 9, 12, 43–4, 50,
    167–9
  general theory of ix–x, 2, 7, 9–14,
    28–9, 42–3, 45–6, 101, 165–75
  identity of 28–29, 43–4, 46–54,
    166–7, 171
abstraction, *see* abstracts, abstraction
    principles, generation
abstraction principles
  acceptability of x, 3–15, 27 n., 28
  analyticity of 29, 171–2
  extensional 3, 11, 23, 42, 45, *see also*
    abstraction principle – law V
  formulation of 1, 3
  genuine 5, 192
  Hume's 2, 8, 15, 17, 19–24, 27,
    31–35, 35–41, 42–6, 73–4, 76, 81,
    83–7, 90–100, 155, 190, 192
  hyper-inflationary 6, 101, 156–61,
    192
  inflationary 4, 5–6, 22, 43–4, 54 n.,
    101, 156–61
  invariant, *see* logical
  law V 2, 3, 4, 35, 39, 41–2, 46, 73,
    137
  logical 7, 43, 45–6, 101, 138–56
  objectual versus conceptual 1–2
  predominantly logical, *see* logical
  predominantly invariant, *see* logical
  stable 10, 13
  tenable 9, 10, 12

analysis (real) ix–x, 2–3, 42–3, 46,
    190–1
analyticity 41, 72, 77, 76–7, 171–2 *see*
    also abstraction principles
    – analyticity of.
arithmetic ix, 2–3, 32–3, 36, 41, 46,
    86, 92, 189–91
axioms of abstraction, *see* abstracts
    – general theory of

Boolos G. 5, 42, 92 n. 22, 127, 138,
    192 n
Brandl J. 89 n.

Caesar Problem 24, 38–9, 44, 56, 60,
    66–7, 68–77, 88–9, 92, 95–6, 100
cardinal number
  as a class, *see* abstracts – as
    equivalence classes
  inaccessible 7, 14
  generalized 14, 43, 47
  Mahlo 13 n. 4
  regular 13 n. 4, 42
  small 7, 42, 45
  unsurpassable 7, 11, 14, 15, 157
categoricity 22, 27, 93, 101,
    122–32, 189
Cantorian set theory 192
completeness rule 70
comprehension principle 12, 45, 103,
    112, 174, 192
conservativeness 191, 192 n.
consistency 35–6, 164
context principle ix, 35, 38,
    46, 55–100, 132, *see also* Caesar
    problem
criterion
  absolute 27, 86, 112–3, 116, 121–2

criterion (*cont.*)
  of identity 3–4, 5, 7, 9, 22 n., 27, 42,
    43, 48–53, 86
  invariant 7–8, 86, *see also*
    abstraction principles – logical
  tenable, *see* tenability

definition 15–35, *see also* categoricity,
  reference – determinate
  via abstraction 20–3, *see also*
    abstraction principles
  circular, *see* impredicative
  contextual, *see* context principle
  creative 56–7
  deterministic 16, 33, 38
  duplicate 71
  essentialist 29–31, 32
  explicit, *see* implicit
  implicit 16, 33, 35–6, 41, 56, 78, 81,
    87 n., 100
  impredicative 27 n.
  inductive 26, 70
  materially effective 33–4
  noncircular, *see* impredicative
  predicative, *see* impredicative
  by reconceptualization 35–41, 57
  recursive 19–20
  referentially effective 17, 34, 38, 56
  referential import 17, 19
  semantic import 18, 19, 33
  simultaneous 16
Drake F. R. 162
Dummett M. 27 n., 35, 39, 55 n.,
  79 n., 89 n., 92 n. 22, 127, 137

equivalence relation (defined) 4
essence, *see* definition – essentialist
explicit definition, *see* definition –
  explicit
existence, *see* abstracts – existence of
extension, *see* principles of abstraction
  – extensional, law V
extensionality
  principle of 54, 107, 109, 174, 189,
    191
  see also abstraction principles
    – extensional

Field H. 25 n., 36, 79 n.
Foreman M. 158
Frege
  on abstraction 1–4
  Begriffsschrift 39 n. 20
  consistency proof 137–8
  Grundgesetze 2, 69 n., 137
  Grundlagen 35, 38, 55, 66, 79 n.,
    109, 110
  definition of number 28–9, 32, 66
  derivation of arithmetic 157, 189,
    191
  generality of logic 109
  numerals 95–8
  object/concept distinction 1, 102
  on reference 67, 79
  on sense 18, 19
  switching argument 22 n., 110
  see also Caesar problem, law V,
    logicism, reference, sense

generation
  of abstracts, 20–1, 26
  of means of abstraction 27–8
  of models 101, 111, 118–22,
    175–89
  of numbers 92
  of understanding 27 n.
grade (of statements) 82

Hale R. 27 n., 35 n., 39 n. 20, 40, 42 n.,
  47 n., 48, 49, 61 n. 2, 68, 74 n. 12,
  92 n. 22
Hazen A. 64 n. 4, 192
Heck R. 5, 10, 41 n
Hodes H. 55 n., 192
holism (referential) 79–80, 87 n
Horwich P. 18 n. 10
Hume's principle, *see* abstraction
  principle – Hume
Humean operator 86, 89
hyper-inflation, *see* abstraction
  principles-hyper-inflationary

identity, *see* abstracts – identity of,
  criterion – of identity, mixed/
  unmixed identities

implicit definition, *see* definition
  – implicit
impredicativity 45, 56, 62, 64 n. 4,
  81–90, 83, 90–100, 137, *see also*
  definition – impredicative.
inflation, *see* abstraction principles
  – inflationary

law V, *see* abstraction principles
  – law V
limited access 76–7
logic second-order 8–9, 11–12, 101–2
  versus ZF, 10–13
logicism 32–3, 38, 41, 46

Martin T. 13 n. 4
Menzel C. 14 n.
mixed/unmixed identities 66, 68,
  88–9

NBG 42 n. 24, 48 n.
nominalism 36, 132
number, *see* abstraction principle
  Hume's, analysis, cardinal
  number, Frege, definition of
  number, definition essentialist.

paradox, *see* Russell's paradox
Parsons C. 41, 55 n.
Parsons T. 110
Peacocke C. 53 n. 28
predicativity, *see* impredicative
predominantly logical, *see* abstraction
  principles – predominantly
  logical
procedural postulation v, 36, 56, 100
postulation, *see* procedural
  postulation

quantification unrestricted 84, 89–90
quasi-Humean 92–3
quasi-numerical 29

Ramsey 25 n.
real number, *see* analysis
reference 1, 16–17, 31, 35, 67, 72,
  76–7
  canonical 78, 80
  definite/indefinite 82–3
  determinate 31–32, 38, 56, 60, 65,
    71, 77–81, 100
  indeterminate, *see* reference-
    determinate
  see also context principle,
    definition, holism
Roman Problem 67, 68, *see also*
  Caesar Problem
Russell's paradox 2–3, 6 n. 2, 164, 167,
  173

Schroeder-Bernstein Theorem 162
Schroeder-Heister P. 110
Shapiro S. 45 n. 23, 163, 192 n
sense 16–19, 33, 35–6, 37, 40, 57–8,
  77
  see also definition, reference
stipulation 35, 56
Sullivan P. 89 n.

tenability 9–10, 12–14, 101, 114–8
Tennant N. 53 n. 28

unrestricted quantification, *see*
  quantification – unrestricted

Weir A. 42 n. 23, 192 n
Woodin W. H. 158
Wright C. 25 n., 27 n., 35, 35 n., 40,
  41, 42 n., 48, 55, 61 n. 2, 64 n. 4,
  68, 74 n. 12, 84 n., 89 n., 92 n. 22,
  95, 96 n., 97 n., 192 n

Zermelo 48
Zermelo–Fraenkel set theory (ZF,
  ZFC) 7, 10, 12, 14, 172

Index of First Occurrence of Formal Symbols and Definitions

L^2 101
L^3 101
purely logical 102
C, D, E 102
R, S, T 102
P, Q, P′, Q′ 102
$L^§$ 102
§ 102
$L^§$ 102
Dm(R) 102
Rg(R) 102
Fld(P) 102
Refl(R) 102
Sym(R) 102
Trans(R) 102
Eq(R) 102
⊆ 102
≡ 102
compl 103
$\rightarrow_R$ 103
$1 - 1(R)$ 103
eq_R 103
eq 103
≤ 103
beq 103
Perm 103
L^2 103
Comprehension 103
L^3 103
Φ 103
T^ϕ 103
$L^§$ 103
T^ϕ 103
identity criterion 103
system of identity criteria 103
abstraction principle 104

logical 104
grounded 104
L-criterion 104
< (precedes) 104
definable over 104
Ab 104
I 104
Nab 104
Uab 104
restricted 104
ϕ_ψ 104
$T^{\phi,\psi}$ 104
$\vdash^2$ ($\vdash^3 \vdash^\psi, \vdash^\Phi$) 105
model 105
§-expansion 105
reduction 105
abstracta 105
concreta 105
pure 105
extensional 106
represented 106
full 106
set-theoretic 106
standard 106
$\mathbf{M}_c$ 106
cardinal abstractor 106
bicardinal abstractor 106
divisor abstractor 106
$\equiv_§$ 106
$\|_§$ 106
$\mathbf{P}_§$ 106
strictly separated 106
separated 106
weakly separated 106
$\mathbf{E}_{\phi,\mathbf{M}}$ 107
criterial relation 107

$P_{\phi, M}$ 107
local set relation 107
global set relation 107
E^{ϕ} 107
extensionality lemma 107
Func 108
Mon 108
$FP_{\phi, B}$ 108
$LFP_{\phi, B}$ 108
least fixed point theorem 108
equivalence lemma 108
inclusion lemma 109
extended extensionality lemma 109
permutation lemma 109
invariance corollary 110
switching lemma 110
§-extensional 111
extensionality principle 111
invariant set-relation 111
outer invariance lemma 111
relativized formula 112
restricted formula 112
M/N 112
partial (restricted) model 112
total model 112
absolute formula 112
absoluteness lemma 113
internal global relation 113
internally (I-) invariant set
    relation 113
tenable criterion 114
$\mathbf{d}$-tenable 114
tenable on $\mathbf{c}$ 114
$\mathbf{d}$-tenable on $\mathbf{c}$ 114
stable criterion 114
stabilize (for an identity
    criterion) 114
generally tenable 114
indefinitely tenable 114
anti-inflation lemma 115
non-inflationary cardinal 116
static 116
partition cardinal 116
(extended) anti-inflation lemma 116
tenable system of identity criteria 116
stable system of identity criteria 117
Eq_{ϕ} 118

Noninfl$_{\phi}$ 117
stability number 118
Hanf number 118
stabilize (for an arbitrary
    sentence) 118
critical ordinal 119
minimal model 119
axiom of constructibility 119
downward-closed 120
upward-closed 120
closed 120
G_K 120
K-minimal 120
Closed(C) 120
Min 120
Absolute(ϕ) 122
Min(C) 122
internally similar 123
extension lemma 123
chain lemma 124
categorical extension lemma 124
categoricity corollary 126
internally similar 127, 128
indistinguishable from 130
quasi-upward closed 130
critical ordinal 131
strictly minimal model 131
individual constants 132
object term 132
relation term 132
concept term 132
rk (rank) 132
objectual term 133
relational term 133
formula (of term language) 133
(term) domain 133
T_C, T_{ϕ} 133
$L(\phi)$ 133
true (for term language) 133–4
complexity 134
$Eq_{=}$ 134
$\approx_T$ 134
$|t|$ 134
regular identity criterion 135
representative domain 136
non-inflationary 138
bicardinal 138

cdstr (cardinality distribution) 138
yields 139
combination 139
representative combination 139
~ (symmetric difference) 139
$\equiv'$ (almost sameness) 139
very different 139
small/large subset 140
almost universal subset 141
bifurcates 141
bifurcator 141
theorem on representative
        combinations (TRC) 142
entails (for cardinality
        distributions) 142
Q, Q_c 143
$[M]^c$ 143
strictly acceptable 143
exponentially small (cardinal or
        subset) 143
exponentially large (cardinal or
        subset) 143
$\approx_0$ (basal relation) 144
as refined (subsumes) 144
characterization theorem 144
strictly I-acceptable 145
$\approx_1$ (super-basal relation) 145
c^+ (successor) 147
$\equiv_d$ 147
Icdstr (internal cardinality
        distribution) 147
numeric 149
CCD (considerations of cardinality
        distribution) 148
$\approx_{num}$ 151
$\leftrightarrow$ (biequivalence) 151
$\leftrightarrow_L$ (left connection) 151
$\leftrightarrow_R$ (right connection) 151
$\|_L$, $\|_R$, 151
P_L, P_R, 151
K-invariance 152
internal K-invariance 152
$\approx_{0,K}$, $\approx_{1,K}$ 152
(internal) K-acceptablility 152
Expsmall, Explarge 153
Stb, Unstb 153
$\approx_2$ 154

cell 156
strictly admissible (cell) 156
C_M (cells) 156
E_M (equivalences) 156
cp (cardinality of precedessors) 156
cf (cofinality) 156
cdstr($\approx$), cdstr(X) 156
unsurpassable (cardinal) 157
IE_M, IC_M 158
(I)-replete (model) 158
K-cell 158
K-admissible 158
cdstr$_K$ (relative cardinality
        distribution) 158
$C_{M;K}$, $E_{M;K}$ 159
(internally) k-replete (model) 159
exbd (exponential bound) 159
predominantly (I)-invariant 159
(broadly) (I)-invariant 159
(broadly) (I)-admissible 159
Abstr (abstraction predicate) 165
Ab (predicate for abstracts) 165
App (applicability predicate) 165
(I)-Inv (invariance predicates) 166
(I)-Inv$_K$ (K-invariance predicate)
        166
(I)-Prdinv (predominant invariance
        predicate) 166
Noninfl (non-inflation predicate)
        166
GA, GA$^+$, GA^2, SGA, SGA^2 170
STE^2, TE 174
steady (cardinal) 175
exponential (cardinal) 176
non-inflating (equivalence) 177
K-acceptable (equivalence) 177
(broadly-, strictly-) acceptable
        equivalence 177
indexed cell 177
K-admissible (cell) 177
(broadly, strictly) admissible
        (cell) 177
expansion (of an equivalence) 178
K-legitimate (expansion) 178
(broadly, strictly) legitimate
        expansion 178
K-admits (a cell) 178

(broadly, strictly) admits 178

finitary-infinitary (indexed cell) 182

(absolutely) indistinguishable
  (indexed cells) 183

$U_{I,\xi}$, $V_{I,\xi}$ 184

introduced (cell) 184

U_I, V_I 186

conservative extension 191